Praise

"Need-to-read inside informatio[n] ... the best source in the business." – Daniel J. Moore, Member, Harris Beach LLP

"The Inside the Minds series is a valuable probe into the thought, perspectives, and techniques of accomplished professionals..." – Chuck Birenbaum, Partner, Thelen Reid & Priest

"Aspatore has tapped into a gold mine of knowledge and expertise ignored by other publishing houses." – Jack Barsky, Managing Director, Information Technology & Chief Information Officer, ConEdison *Solutions*

"Unlike any other publisher – actual authors that are on the front-lines of what is happening in industry." – Paul A. Sellers, Executive Director, National Sales, Fleet and Remarketing, Hyundai Motor America

"A snapshot of everything you need..." – Charles Koob, Co-Head of Litigation Department, Simpson Thacher & Bartlet

"Everything good books should be - honest, informative, inspiring, and incredibly well-written." – Patti D. Hill, President, BlabberMouth PR

"Great information for both novices and experts." – Patrick Ennis, Partner, ARCH Venture Partners

"A rare peek behind the curtains and into the minds of the industry's best." – Brandon Baum, Partner, Cooley Godward

"Intensely personal, practical advice from seasoned dealmakers." – Mary Ann Jorgenson, Coordinator of Business Practice Area, Squire, Sanders & Dempsey

"Great practical advice and thoughtful insights." – Mark Gruhin, Partner, Schmeltzer, Aptaker & Shepard PC

"Reading about real-world strategies from real working people beats the typical business book hands down." – Andrew Ceccon, Chief Marketing Officer, OnlineBenefits Inc.

"Books of this publisher are syntheses of actual experiences of real-life, hands-on, front-line leaders--no academic or theoretical nonsense here. Comprehensive, tightly organized, yet nonetheless motivational!" – Lac V. Tran, Sr. Vice President, CIO and Associate Dean Rush University Medical Center

"Aspatore is unlike other publishers...books feature cutting-edge information provided by top executives working on the front-line of an industry." – Debra Reisenthel, President and CEO, Novasys Medical Inc.

www.Aspatore.com

Aspatore Books, a Thomson Business, is the largest and most exclusive publisher of C-level executives (CEO, CFO, CTO, CMO, partner) from the world's most respected companies and law firms. Aspatore annually publishes a select group of C-level executives from the Global 1,000, top 250 law firms (partners and chairs), and other leading companies of all sizes. C-Level Business Intelligence™, as conceptualized and developed by Aspatore Books, provides professionals of all levels with proven business intelligence from industry insiders—direct and unfiltered insight from those who know it best—as opposed to third-party accounts offered by unknown authors and analysts. Aspatore Books is committed to publishing an innovative line of business and legal books, those which lay forth principles and offer insights that, when employed, can have a direct financial impact on the reader's business objectives, whatever they may be. In essence, Aspatore publishes critical tools—need-to-read as opposed to nice-to-read books—for all business professionals.

Inside the Minds

The critically acclaimed *Inside the Minds* series provides readers of all levels with proven business intelligence from C-level executives (CEO, CFO, CTO, CMO, partner) from the world's most respected companies. Each chapter is comparable to a white paper or essay and is a future-oriented look at where an industry/profession/topic is heading and the most important issues for future success. Each author has been carefully chosen through an exhaustive selection process by the *Inside the Minds* editorial board to write a chapter for this book. *Inside the Minds* was conceived in order to give readers actual insights into the leading minds of business executives worldwide. Because so few books or other publications are actually written by executives in industry, *Inside the Minds* presents an unprecedented look at various industries and professions never before available.

INSIDE THE MINDS

The Changing Landscape of Natural Disaster Relief

Government Officials on Allocating Financial Resources, Organizing Outreach, and Managing Relief Programs

Mat #40740787

BOOK & ARTICLE IDEA SUBMISSIONS

If you are a C-Level executive, senior lawyer, or venture capitalist interested in submitting a book or article idea to the Aspatore editorial board for review, please e-mail TLR.AspatoreAuthors@thomson.com. Aspatore is especially looking for highly specific ideas that would have a direct financial impact on behalf of a reader. Completed publications can range from 2 to 2,000 pages. Include your book/article idea, biography, and any additional pertinent information.

Inside the Minds Project Manager, Kristen Skarupa; edited by Jo Alice Darden; proofread by L. Jones

ISBN 978-0-314-99176-8

For corrections, updates, comments or any other inquiries please e-mail TLR.AspatoreEditorial@thomson.com.

First Printing, 2008
10 9 8 7 6 5 4 3 2 1

CONTENTS

Changing the Victim Mindset in Natural Disaster Relief

W. Craig Fugate

Director

Florida Division of Emergency Management

The Present Situation of Natural Disaster Relief

Natural disaster relief encompasses a wide variety of involvement and duties. There are broad groups involved, such as individuals, the private sector, which runs the spectrum of companies, from those that do disaster work to the businesses and non-profits in the community that provide services every day. Then you have volunteer organizations, individual volunteers, and finally, government, in which local, state, and federal agencies have disaster relief roles.

After Hurricane Andrew struck Florida in 1992, we had a response that was a public nightmare. We did not do well as a state, and the federal government took a black eye. Our response to that event was to address many shortcomings by building a system heavily based upon government. To many degrees, we improved government response tremendously. We built a robust volunteer system through the standing volunteer agencies, such as the Salvation Army and Red Cross, getting them to recognize that they could not continue to operate without each other. We got them to sit down in our state EOC (Emergency Operations Center) and work as a team, versus working independently. All these agencies have specific strengths and weaknesses, so why should each try to be all things to all people? We asked them to work together to maximize their strengths, and if there were other agencies stronger than they in other areas, to get them involved. This is not about who will get the best press; it is about meeting the needs of the victims. If we do that, we will all get good press.

We have built a good system of partnerships from the local level to the state level to the federal level. We have been looking at how to leverage resources that are not directly affected through mutual aid, both within the state and then with neighboring states through the EMAC (Emergency Management Assistance Compact), which is our state-to-state mutual aid agreement, and then working with FEMA (the Federal Emergency Management Agency).

As explained on the EMAC Web site, located at www.emacweb.org:

> The Emergency Management Assistance Compact (EMAC), established in 1996, has weathered the storm when tested and

> stands today as the cornerstone of mutual aid. The EMAC mutual aid agreement and partnership between member states exist because from hurricanes to earthquakes, wildfires to toxic waste spills, and terrorist attacks to biological and chemical incidents, all states share a common enemy: the threat of disaster.
>
> Since being ratified by Congress and signed into law, in 1996 (Public Law 104-321), 50 states, the District of Columbia, Puerto Rico, Guam, and the US Virgin Islands have enacted legislation to become members of EMAC. EMAC is the first national disaster-relief compact since the Civil Defense and Disaster Compact of 1950 to be ratified by Congress.

In 2005, Hurricane Wilma was so big the system got stretched tightly. We discovered we had been ignoring our private-sector retailers. When a hurricane hit and the power went out, usually all the grocery stores closed. The retail sector realized they could not afford to be out of business for long periods, particularly if the competition got a generator and opened their store. You did not want to be the retailer who was not open. We actually found ourselves in 2005 handing out free food, water, and ice in the parking lot of an open grocery store. We found that as bad as the reports were during Hurricane Wilma, with two to three days to get things up and running because a huge geographical area was affected, on the day after the hurricane, only about five of the major grocery stores in that entire area were not open.

We had to learn to quit competing with the private sector. We met with them and came up with this realization: the best way for us to manage getting commodities into a community is to quit competing with and duplicating the private sector. We needed to bring them on board as part of the team, and if they were up and running in those areas, then we do not need to be handing supplies out in that area. We needed to know which areas they were not in. (They tend not to be in the inner-city, low-income, depressed areas, or in rural areas of the state.) Those are the areas we should have gone to first, but we actually got to last because we were trying to meet the population needs of the 80 percent versus the other 20 percent who were not serviced by those stores and the people who did not have the ability to go and get supplies, such as the disabled and elderly.

The private sector on the retail side was providing these services every day in a community. Why did we make the assumption that all of that was going to go away and they would not try to open? We now sit down and routinely bring the retail sector in on our exercises. When we do conference calls during storms, they are on the calls. We established it this way: if you have your stores open, then we are not going to do mass distribution in those areas; we will go where you are not. If your stores are shut down or destroyed, we will fill in those gaps until you come online, and we will share that information freely in a disaster and work as a team. That, to me, has been a huge piece in recognizing that, particularly in the retail sectors, we were undervaluing what they were doing, and often being oblivious to what they were doing.

It goes back to the question of where does government have a role, and where does the private sector have a role? Our most important task will be life safety missions; making sure areas are safe and secure; making sure we can get to the injured; and making sure there are enough medical capabilities brought back online so that we can at least treat and stabilize people in the area of impact, even if we have to send them out of that area for further care. Then we need to start addressing those immediate needs that the private sector cannot meet, or is not meeting, for people who do not have the money or ability.

That gets into deciding where we send supplies and where we set up shelters. The next step is getting communities back on their feet. You have to get schools open, even if you do not have housing for people. If you do not get schools open, your recovery is dead. This is an interesting phenomenon that parents understand. Parents will not risk their children's future. If they do not see the public or private education systems their children are attending coming back online quickly, if it will be months before it happens, they will move. They will take their kids where education is offered. If you do not have schools, you cannot get the recovery going.

Tulane University was nearly wiped out by Hurricane Katrina, and is probably one of the most under-told stories of how successfully they faced adversity and made the decision to come back. They found early in the process that they would be able to get the faculty and staff back, but they could not depend upon Orleans Parish getting the schools open. They ended up going into the charter school business because people would not come back if their kids could not go to school. In many of our communities hit in the hurricanes of 2004, people remarked time and again, when they saw the school buses for the first time, three weeks after Hurricane Charlie leveled Charlotte County and destroyed seven of the fourteen schools, that this was the first time they really believed the schools would come back.

Opening schools is a psychological boost, and so important to a community. One of the fundamental things is education and getting it back online. Putting a lot of emphasis on that, trying to move communities through the trauma of a disaster, and getting things like schools open really stabilizes the community. The community can then start to rebuild. If we do not do those things, if you just get a community stabilized by only addressing housing, for example; if you do not start bringing things such as schools back online, then people start leaving. Your workforce is gone, the demand for consumption is gone, and then it gets dicey whether you can really get a community back online. If there was not something intrinsically there that will regenerate the economy, communities will not come back.

In Mississippi, after Hurricane Katrina, officials made the strategic decision to get rid of their rule that said casinos had to float. They decided to be pragmatic about this and allow casinos to come onshore. If they did not have gambling on the coast of Mississippi, south Mississippi would still be further behind. If New Orleans was not a port and if the French Quarter had not been relatively spared in Katrina, then New Orleans would not be coming back. New Orleans is fortunate that the Garden District, the French Quarter, and the central business district, though badly damaged, were recoverable. If they had lost those tourist destinations, then the only thing New Orleans had going for it would have been the port itself, which would not have sustained any sizeable recovery of the community.

The biggest issues right now in natural disaster relief are the disparity between the public's expectation of what should happen and the reality of what our current government-based response system looks like, and the inability to translate programs based upon reimbursement into what the public thinks should be happening.

For example, FEMA has become a noun for emergency management. It stands for what I do in the state of Florida, but people seem to think that FEMA is who responds to disasters. FEMA is not that agency. They are an agency that administers reimbursement programs if a disaster meets a certain threshold the president declares, based on the direction given by Congress. This creates tension with the public during a disaster, because FEMA, despite all this national recognition, cannot do what seems like the logical thing to do.

I think over the last twenty years, Congress, different administrations, and state and local governments have designed and built this system around the Stafford Act (Robert T. Stafford Disaster Relief and Emergency Assistance Act, Public Law 93-288, as amended, 42 U.S.C. 5121-5207, and Related Authorities. UNITED STATES CODE TITLE 42. THE PUBLIC HEALTH AND WELFARE CHAPTER 68. DISASTER RELIEF), which is a reimbursement program. After Katrina and other large-scale, catastrophic disasters, we learned that this system does not work. The system consists of local and state governments responding and expending funds, waiting for reimbursement by the federal government, and then having the federal government determine what was eligible and not eligible. In future disasters, decision-makers remember what was not eligible, so they decide not to do that because they will not get reimbursed.

Whether we will get reimbursed is not the right question to ask during a natural disaster. The right question should be: What do you need to do to protect your citizens? What do you need to do to meet the immediate needs to stabilize the situation and move forward? We have ended up defining how we respond to large-scale and catastrophic disasters by reimbursement programs, and I think we need to turn this around and define how we need to respond to large-scale disasters by determining the things we need to do to get communities stabilized quickly and meet basic challenges in very short periods.

One of the challenges I see in the world of natural hazards, terrorist attacks, nuclear power plant failures, and chemical emergencies is the response issues. The same basic things need to happen; just the scale needs to change. In a disaster, whatever the cause, if you have injuries, if you have people trapped, you have to complete search and rescue quickly—not in days to weeks, but in hours to days. You cannot do that if the area is not secure, so you have to get security in there quickly. You have to be able to physically get into areas—for example, when the I-10 bridge failed outside Pensacola, Florida, during Hurricane Ivan.

Our model for emergency management in the United States is based upon the lowest level of government having primary responsibility, and the state and federal government will respond with assistance as needed. The reimbursement model drives this, and local governments need to exceed and expend their local resources and their local mutual aid before they go to the next level. They have to assess and determine how bad it is, and then send their requests to the state. We are supposed to review it and determine where the resources are and send those resources in. If the state lacks the resources or capabilities, they make their request to the federal government through FEMA. That means it takes about seventy-two hours to start getting enough critical mass on the ground to stabilize the situation, but in seventy-two hours, you can lose control of the situation. In very large-scale catastrophic disasters, you can come to a screeching halt.

Look at Katrina. There were a lot of resources sitting right on the perimeter of New Orleans that could not go in because of the perception of lawlessness, and these unarmed volunteers and professional responders were not going to be sent into an area where there was a possibility that they might be at risk. Until an area is secured, most resources are frozen. What we have been pushing to shift in Florida is that if I know I have something bad happening, I am not going to use the reimbursement model, assess it, look at what I need, try to right-size it, and get everything in there at the lowest cost. I need to focus in on the things I need to do in the next twenty-four hours. I need to make it secure and safe so people have a sense that they are not by themselves. In most cases, lawlessness and looting are overblown and over-reported, but the perception can stop you in your tracks during a response. I need to have a lot of search and rescue and medical capabilities moving in, and I cannot wait to find out if the situation is bad to do that. In a hurricane, I need to move things before the storm ever hits.

Whether I am a local or state government, factoring in who will pay for this, whether we can afford to do this, and whether FEMA will reimburse us on it goes back to an economic model that says unless it is reimbursable, we will not make those decisions early enough so we can get a community stabilized, so we can start moving to recovery, and we will lose momentum. Go talk to people in the mental and behavior health communities and ask how people deal with the trauma of disaster, what things you need to do to get communities where they can start the recovery process. What kind of mindset is needed? The first thing people need is to feel safe, and until they feel safe, they cannot even take the next mental step of dealing with how catastrophic and upsetting this situation is. Why am I spending all this time and energy on having to have intelligence and all the data to know how bad it is? Why not go ahead and mobilize the resources as if it is going to be bad? If it is not, it will be done more quickly. If it is, and the first thing people see is a National Guard HumVee going down the street, that all by itself has not changed a lot for them, but it has told them there is security, and they are not by themselves.

Here are my standing orders for a hurricane response:

1. Establish communication with affected areas
2. Search and rescue security
3. Meet basic human needs
    - Medical
    - Water
    - Food
    - Shelter
    - Emergency fuel
    - Ice is a distant sixth (unless it's really hot)
4. Restore critical infrastructure
5. Open schools/local businesses
6. Begin the recovery

I am trying to get people to think more along the lines of the perspective of your family and community and what things need to happen to get you through the first critical hours. Focus on lifesaving, life-safety, and life-sustaining issues. Do not get hung up on temporary housing, picking up the debris, and fixing everything. My focus is on keeping people alive. I cannot afford to lose lives because my response is based upon a reimbursement model.

Once I have gotten to where we are not losing any more people to the event, whether it is a collapsed structure or a tanker fire—once I have gotten to the point where nobody else is dying or being injured by the event, then I can go to the next step. Do I have the ability to meet the very basic needs of shelter, medical care, water, food, and essential items? Again, I am not trying to solve every problem up front, but I am trying to make sure everyone we *can* get through as a survivor survives.

The next step is whether we can meet their basic needs. Then we start moving them into the process of engaging in this disaster as owners of that response, versus everyone coming in and taking care of them. You hear a lot in a disaster that everyone is a victim, and when you talk to behaviorists, what we are really doing is creating a dependency. The reality needs to be victims are the ones we lost in the initial event. Everyone else is a survivor.

Disaster behavioral health early intervention:

1. In a disaster, survivors experience risk: *Move them to safety.*
2. In a disaster, survivors experience fear and distress: *Provide support for calming.*
3. In a disaster, survivors experience separation from loved ones: *Help re-establish connectedness.*
4. In a disaster, survivors experience helplessness: *Rebuild self-efficacy.*
5. In a disaster, survivors experience despair: *Rekindle hope.*

We have a tendency to build these government-based systems and forget that the community itself is the one that will have to deal with this. We can help, but you have to get people to take ownership and give them control of the situation. Self-efficacy is a great way to get populations engaged in their community, take ownership, move them to recovery, and get some sustainability; then the communities will come back.

We have essentially been doing what I call nickel-and-dime disasters for a long time under a reimbursement program. That is a good way to do business until you get to these very large, complex disasters, and then it is hard to change to a system that cannot wait for the answers, that cannot be based upon a reimbursement model, and cannot be based upon a government-centric approach to dealing with disasters. That cuts across everything, whether it is terrorism, a pandemic, or a hurricane. It is a shift that, as policymakers, we really have to get back to. Are we doing this because someone else is going to pay for it, or are we doing it because this is what we need to do to get our community back on its feet?

The Legal Aspects

We are a republic, and part of our constitution vests most domestic issues in the hands of the states and the governors, not the Congress and executive branch, which are often seen as the big players. The reality is the Constitution vested power for domestic issues, including response to disasters, in the states. Each state has differing legislation, and based upon their history of disasters, biases, fears, or concerns about the excessive powers of government, they have a variety of authorities.

For example, up until a few years ago in Texas, the governor did not have the authority to order evacuation—only the local county judges and city commissioners did—so they could not move very well at the state level to facilitate an evacuation beyond what the local officials wanted to do. Other states do not give their governors much latitude in budgeting for emergency appropriations because they are fearful of abuses by governors.

Most of these laws came into effect about the Civil Defense Act of 1950, dealing with the cold war and nuclear attack scenarios. Some states, such as Florida, because we were right in the middle of the Cuban Missile Crisis, gave their governors extraordinary powers in disasters to take action (Florida Statutes Chapter 252). many people want to know why you do something or cannot do something, and unless you have the legal basis to act or have the legal authority to do it, you may be constrained. There may be things in some states that governors can do, that in other states only the local governments can do. Not every state has the same authorities and responsibilities vested the same way in every government.

When you look at national news, there is a tendency that everything should look, feel, and taste the same, and that we should have a McDonald's form of government in all fifty states, but we do not. There are fifty different states of laws and histories. They have had experiences with different disasters, and that has driven how legislatures have crafted and defined what emergency powers they have, what day-to-day powers they have, and where those powers are vested or shared at the state and local levels.

Then you get to the federal level of government, and as discussed earlier, most of the FEMA programs are driven financially, by what they pay for. The public perception, because of the way the national news portrays it, is that when FEMA comes in, they are in charge, or that the president is in charge. That is not the case. It is the governor who is in charge, unless the president declares that there is a state of insurrection, or that the governor in that state is incapacitated, and there is no Constitutional government, or they are failing to protect the civil rights of their citizens. Those are the only times that the president would be in a situation to take charge of a disaster.

There is actually good case law for this. The attorney general and the president authorized and federalized the National Guard during school integration because the governors were failing to carry out and protect the civil rights of their citizens. During the Los Angeles riots, the governor turned to the president and said it was beyond state and local control, and the president declared a state of insurrection was occurring in California, federalized the National Guard, and brought in the active duty military to support them in dealing with the aftermath of that riot. There was a legal basis to do it.

Many people thought that President George W. Bush should have taken over the response of Katrina and that someone should be put in charge of it. The problem is that we are a republic with a federal system of government, which I think is different from people's perception of our government. I think most people think we are nationalist government, where the president and Congress have day-to-day governing responsibility over every citizen. That is not the case. If the governor asks for help, FEMA and the federal response can provide it in certain situations and under certain declarations. They can, in many cases under certain situations, and at the governor's request to the president, declare a disaster area and provide financial reimbursement for various costs that Congress has authorized, but they are not in charge unless you are under a state of war or a state of insurrection, where a state government is incapacitated.

In Katrina, the president would have to have demonstrated that the governor of Louisiana was failing to protect the civil rights of her citizens, but that would have been a political decision. The governor is elected by the state's citizens, is held accountable to them, and should be judged by the electorate. As a republic and as a federalist government, those powers that are not explicitly reserved for the federal government of the Constitution are expressly given to the states to govern themselves. If you do not understand that, then you cannot understand why FEMA does not just take over. FEMA was not built and designed as a primary response agency to come in and run a disaster, but rather was charged by Congress and the president and under the Stafford Act to be the federal agency that coordinates, on behalf of a governor's request, the federal government support to a state, and to administer the reimbursement and funding programs of the Stafford Act.

When you look at the Stafford Act, you need attorneys to help you write declaration requests, deal with that process, and deal with appeals and everything else in supporting that process. On the other side of it, when governors respond to a disaster, unless they have experienced staff who have worked with a lot of these disasters, they generally find themselves—particularly as you change administrations—not always sure of what they can and cannot do. So their legal staffs have to make sure they have a good understanding of that state's emergency management laws and what the government can and cannot do.

Many times, governors will do something, and then the legal staff must figure out whether they had the authority to do it. That is part of what our legal team does for us. We have a full-time attorney assigned to our team who is our adviser. You sometimes reach the question: Can we do this—do we have the legal basis to act? If we do not, what do we need to do to address this problem? Because all fifty states have differing ways of approaching disasters, it is important that the first thing governors do before they are sown into office is make sure their legal advisers and staff understand and can implement the state's emergency management authorities for a governor in a time of a crisis. They also need to be able to provide the recommendations and policy guidance when you hit those gray areas that are not clearly spelled out. Many statutes have language in them that allows them to be interpreted as broadly as necessary. For a response-geared organization, that means unless you are told no, you are going to do it. However, this can put you in a problematic situation.

For example, in 1998, during the wildfires in Florida, the late Governor Chiles made a decision based upon recommendations from various agencies to ban the sale of fireworks in the state. We first were challenged: Did the governor have the authority to do that under the state emergency management act during a declared emergency? That went through the system, and the appellate courts found that yes, the governor did have the authority. The fireworks distributors' attorneys decided not to challenge the governor's authority under what are called police powers, but they filed a suit against the state on a takings claim that because we prohibited the sale of fireworks, we unlawfully took, without compensation, the ability of the fireworks industry to have earned a return on their investment. Takings claims are a gray slippery slope that I think we will see more and more.

We have to take actions in emergencies, which may mean we will have to go onto your property. There may be use of your facilities; there may be decisions made that adversely affect your property, such as if we order an evacuation, and we order all the hotels and motels in that evacuation area to shut down. In most cases, people say yes, the governor has the right to do that, but you took my income, and I want compensation.

An interesting thing we are seeing in this fireworks case is that it is one thing for me to take and condemn your property and then continue to use it after a disaster, but if I am doing it only temporarily to protect life and property, what is the responsibility of government to the owner of that property, particularly for things like loss of income? I cannot think of any disaster, where someone is affected economically by actions that government took above and beyond what the disaster was, that could not be considered affecting my income. If this is allowed to go to the point where they find for the plaintiffs that there was a taking, and government owes, it will definitely change many decisions people make. It will come back to this: If I make this decision, how much will it cost me, and how much exposure do I have?

Many states try to address these issues up front and use the constitutional authorities the states have to protect life and property, which can include seizing property, ordering evacuations, ordering curfews, and prohibiting the sale of certain items during certain events, and provide broad immunity from civil and criminal penalties for those actions if they are done in that situation. Over the years, there has been push-back against sovereign immunity in the states and local governments to provide more access for citizens to address their grievances on other matters, and it may have, in some cases, weakened that immunity in disasters and opened Pandora's Box. Now everything is subject to a civil dispute process if people feel that they were aggrieved and have some compensation owed to them.

The reimbursement-based model of disaster response has never anticipated or considered the civil suits for the actions taken. There is a big difference between actions taken to protect property that were successful, and things that were done as neglect, or things that were failure to respond, failure to serve, or failure to follow your plans. Ordering evacuation for a coastal community threatened by a hurricane is the rational thing to do, but now if takings claims are being filed, if the states do not address this and provide for that immunity, then that could temper the decision when a situation is not as clear-cut as Katrina heading right at your house.

When is it too late to make the decision? A good example is Hurricane Andrew. On the Friday before it hit, it was a Category 1 hurricane, and Miami-Dade County had to start making decisions. On Saturday, they had to order an evacuation, and the storm did not hit until Monday. There was just about as much chance that Hurricane Andrew could have missed Miami-Dade County as hit it at that point. Probably more than three-quarters of the time, you order an evacuation for a hurricane, and there is no significant damage to those areas that you ordered evacuated. From the legal standpoint of the government, you have to understand what your statutes say you can and cannot do, what your emergency powers are, and the processes of doing that. You need to understand the Stafford Act, how to request assistance, and what that process is. Then you have to deal with the potential consequences of negligence and failure to do the job, as well as the unintended consequences of actions taken to protect citizens that may have resulted in impacts to individuals and business owners who may then seek redress through a takings claim.

Impact on All Levels of the Community

For a long time, we as a nation have talked about disability rights and accessibility issues. The Americans with Disabilities Act (ADA) has been a big challenge to get into the mainstream, and accept that if you are a public venue, you can no longer ignore that people have to be able to get in and out and have accessibility. The ADA seems too hard to implement in a disaster, so we have often ignored the part of our citizenry who cannot access services needed without assistance. I think we are seeing now that the Department of Justice is saying that ADA is as applicable during a disaster as it is during any other time. In fact, it is probably more important during a disaster because people with disabilities find themselves not receiving warnings or not being able to go to shelters or get services because they have a disability and nobody thought to accommodate them.

This goes back to the heart of how disasters affect communities. Our role in the government is to address the needs of our most vulnerable citizens, and yet we have a tendency with ADA to not aggressively address that issue and make sure that we can meet the needs of all of our citizens. Under ADA and under our moral responsibility, I think we have to. These are our vulnerable citizens, so why are we not putting that emphasis on them at a higher level than many other folks who, in many cases, can take care of themselves and have that ability? We tend to look at our disaster response design as one-size-fits-all, which will not work for our elderly and people who have different types of disabilities and our poor.

You could say Katrina was one massive civil rights violation. It goes back to the reimbursement model. We have become molded to complacency, dealing with a lot of small and medium-sized disasters that were handled well locally and on a state response basis. FEMA is primarily a reimbursement agency, and the people who did well were the middle class to the wealthy. The people who did not do well were the poor and the people who were most vulnerable, such as mothers and children; low-income, working families with two or three jobs; and people who thought they were middle class. In reality, they were the working poor, with no backup and no resources, one paycheck away from desolation; the elderly who did not have nearby family, on a fixed income, with no resources; people with disabilities, who disproportionately are poor; and people with mental illnesses and dependence issues. Most previous disasters were never so big that they really showed how vulnerable these people were, and how poorly we were meeting those needs. Katrina just ripped away that façade.

We built a system on reimbursement for the middle class and wealthy, and that was not the challenge. The challenge was that the people who needed the most help were being least served by these programs, and even though we said we do not discriminate based on race, religion, age, creed, etc., we as a whole had built an emergency management system around the middle class. It was based on people who had insurance and mobility and could carry many of these functions. The question then tended to be, "Is this reimbursable? If it is, we will do it, and if not, we won't." A new system needs to be based on what the survivors need to get on with life and recover, and how we can best protect our citizens, and not make the assumption that everyone has a high school to college education, speaks and understands English, receives information the same way, and is able to understand information the same way.

We need to quit being a nation of victims expecting someone else to take care of us. For about 85 percent to 95 percent of the population at any given time, it is the failure to act and take responsibility that takes away from the most vulnerable citizens. I think it has to come back into the perspective that each of us has the responsibility to prepare our families and ourselves to the best of our abilities and not look to someone else to take care of us in those first couple of days after a disaster hits.

The reality is there will never be enough resources to get to everyone at the same time. If we continue with the mentality that we will all be victims, we will let someone else be responsible for us, and we will let someone else come to rescue us, we will still lose people. What we need is a system where we make sure that everybody understands they have a role. I may not be able to do everything, but the things I can do, I need to do. If I can, I will take care of my family and me, but I also need to help my neighbors. If they are safe, then maybe my neighborhood, and if that is safe, then I need to help my community. Disasters should not be the time where we wait for someone else to step in. We need to step up and be ready when disaster strikes. That is a mindset we as a nation have lost.

We have put too much emphasis on the government as the solution, and we fail to see our greatest resources are our neighbors and ourselves, to take care of those things, and then to help others. Look at the World Trade Center. Who took the responders to the hospitals in the first minutes after the collapse? It was not an organized response—it was pedestrians and cab drivers and anybody who could help get those people out of that initial collapse and get them to the hospital. In an earthquake or a major hurricane, we may not see that first responder, that ambulance or fire truck, for a day or two, but we can help each other. We need to take responsibility to learn about hazards and about the things our community faces.

We as a nation need to move from being victims to being survivors. We need to prepare for and be aware of the hazards our communities face. We need to be ready to take care of each other and recognize that we have to go beyond our doorsteps. If we can do those things, I think we can move away from a system that is based upon a reimbursement program. Hopefully, disasters are few and far between, but when they occur, the community needs to respond—the retailers, the individuals, the citizen groups, the governments, the volunteer agencies—because if we wait for someone to come from the outside, they may not get there in time or come at all.

As the director of the Florida Division of Emergency Management, W. Craig Fugate oversees an agency with 138 full-time staff and budget of $1.1 billion. The Florida Division of Emergency Management coordinates disaster response, recovery, preparedness, and mitigation efforts with each of the state's sixty-seven counties and local governments. In September 2003, the Florida Emergency Management Program became the first statewide emergency management program in the nation to receive full accreditation from the Emergency Management Accreditation Program.

From volunteer firefighter, paramedic, and lieutenant with Alachua County Fire Rescue, Mr. Fugate's career in emergency services includes serving for ten years as the emergency manager for Alachua County, Florida.

In May of 1997, Mr. Fugate was appointed chief of the Bureau of Preparedness and Response with the Florida Division of Emergency Management. Over the next four years, the State of Florida faced numerous disasters while continuing to build a nationally recognized emergency management program. In his role as the chief of the State Emergency Response Team, Mr. Fugate saw extensive action in 1998. That year, Florida experienced floods, tornadoes, wildfires, and Hurricane Georges, resulting in more than 200 days of activation for the State Emergency Response Team.

In October of 2001, Governor Jeb Bush appointed Mr. Fugate to serve as director of the Florida Division of Emergency Management.

In 2004, the state saw the largest response to disasters in Florida history, with one tropical storm and four hurricanes—Charlie, Frances, Ivan, and Jean—affecting the state.

Hurricanes set records again in 2005, when Florida was affected by four hurricanes, three of which made landfall (Dennis, Katrina, and Wilma), and two tropical storms. Hurricane Katrina became the most costly natural disaster during its second landfall on the Louisiana and Mississippi Coast. Florida's State Emergency Response Team launched the state's largest mutual aid response under the Emergency Management Assistance Compact (EMAC) in support of the affected states. Hurricane Wilma made landfall on October 24 on Florida's southwest coast, becoming the third costliest storm in U.S. history.

Governor Charlie Crist reappointed Director Fugate to his administration, the role he continues today.

Disaster Preparedness: The Responsibility Is Ours

H. Douglas Hoell Jr.

Director

North Carolina Division of Emergency Management

Components of Natural Disaster Relief

The process of managing natural disaster relief programs and efforts is a partnership between the affected state or states and the federal government (FEMA). FEMA (the Federal Emergency Management Agency) delivers a federal coordinating officer (FCO), who is in charge of administering the federal share of the programs. He brings with him a staff of federal personnel responsible for managing all of the federal programs. The state delivers a state coordinating officer (SCO), who serves as an equal partner to the FCO. The SCO also brings professional state staff to match up with their federal counterparts to manage the disaster relief programs.

The federal and state teams are tailored to fit the needs of the disaster. The larger the disaster, the greater the need will be for more personnel to manage the disaster relief programs. Ultimately, the president of the United States and the governor of the affected state are the responsible parties. Employees of the federal and the state governments carry out the actions necessary to implement the disaster relief programs, and success or failure will reflect directly on the top leadership at the state and federal levels of government.

The insurance industry should be at the top of the list when it comes to agencies or groups involved in natural disaster relief. The responsibility is ours to protect our property against disaster loss through the purchase and maintenance adequate hazard insurance.

Government and volunteer agencies involved in natural disaster relief include the Department of Homeland Security/FEMA, state emergency management, the U.S. Army/Corps of Engineers, the Small Business Administration, Volunteer Organizations Active in Disaster (VOAD), U.S. Coast Guard, Department of Defense, National Guard, public health, social services, mental health, environmental agencies, emergency response agencies, Department of Agriculture, Forestry, information technology, and others. Insurance goes back to the individual's responsibility to prepare for disaster and not expect the government to make things whole following an event. Government agencies at all levels make up the emergency support functions addressed in most emergency and disaster response plans.

FEMA's role in partnership with the state emergency management agency is to administer the Robert T. Stafford Disaster Relief and Emergency Assistance Act. Other local, state, and federal agencies participate in the implementation of the Stafford Act, and they carry out the emergency support functions:

ESF –1 Transportation
ESF – 2 Communications
ESF – 3 Public Works and Engineering
ESF – 4 Firefighting
ESF – 5 Emergency Management
ESF – 6 Mass Care
ESF –7 Resource Support
ESF – 8 Public Health and Medical Services
ESF – 9 Search and Rescue
ESF – 10 Oil and Hazardous Materials Response
ESF – 11 Agriculture and Natural Resources
ESF – 12 Energy
ESF –13 Public Safety and Security
ESF – 14 Long-Term Community Recovery
ESF – 15 External Affairs

Following is a list, taken from FEMA's Web site, of the major disaster declarations nationwide for the calendar year 2007:

Major Disaster Declarations

Number	Date	State	Title
1736	12/27	Missouri	Severe Winter Storms
1735	12/18	Oklahoma	Severe Winter Storms
1734	12/08	Washington	Severe Storms, Flooding, Landslides, and Mudslides
1733	12/08	Oregon	Severe Storms, Flooding, Landslides, and Mudslides
1732	11/30	Indiana	Severe Storms and Flooding
1731	10/24	California	Wildfires
1730	10/02	Texas	Tropical Storm Erin

1729	09/25	Illinois	Severe Storms and Flooding
1728	09/21	Missouri	Severe Storms and Flooding
1727	09/14	Iowa	Severe Storms and Flooding
1726	09/07	North Dakota	Severe Storms and a Tornado
1725	09/07	North Dakota	Severe Storms and Tornadoes
1724	08/31	New York	Severe Storms, Flooding, and Tornado
1723	08/31	Oklahoma	Severe Storms, Flooding, and Tornadoes
1722	08/30	Illinois	Severe Storms and Flooding
1721	08/29	Nebraska	Severe Storms and Flooding
1720	08/27	Ohio	Severe Storms, Flooding, and Tornadoes
1719	08/26	Wisconsin	Severe Storms and Flooding
1718	08/24	Oklahoma	Severe Storms, Tornadoes, and Flooding
1717	08/23	Minnesota	Severe Storms and Flooding
1716	08/08	Maine	Severe Storms and Flooding
1715	08/03	Vermont	Severe Storms and Flooding
1714	07/24	Nebraska	Severe Storms and Flooding
1713	07/17	North Dakota	Severe Storms and Flooding
1712	07/07	Oklahoma	Severe Storms, Flooding, and Tornadoes
1711	07/02	Kansas	Severe Storms and Flooding
1710	07/02	New York	Severe Storms and Flooding
1709	06/29	Texas	Severe Storms, Tornadoes, and Flooding
1708	06/11	Missouri	Severe Storms and Flooding
1707	06/07	Oklahoma	Severe Storms, Tornadoes, and Flooding
1706	06/06	Nebraska	Severe Storms, Flooding, and Tornadoes
1705	05/25	Iowa	Severe Storms, Flooding, and Tornadoes
1704	05/25	Rhode Island	Severe Storms and Inland and Coastal Flooding
1703	05/25	Kentucky	Severe Storms, Flooding, Mudslides, and Rockslides
1702	05/22	South Dakota	Severe Storms, Tornadoes, and Flooding
1701	05/16	Massachusetts	Severe Storms and Inland and Coastal Flooding

1700	05/11	Connecticut	Severe Storms and Flooding
1699	05/06	Kansas	Severe Storms, Tornadoes, and Flooding
1698	05/04	Vermont	Severe Storms and Flooding
1697	05/01	Texas	Severe Storms and Tornadoes
1696	05/01	West Virginia	Severe Storms, Flooding, Landslides, and Mudslides
1695	04/27	New Hampshire	Severe Storms and Flooding
1694	04/26	New Jersey	Severe Storms and Inland and Coastal Flooding
1693	04/25	Maine	Severe Storms and Inland and Coastal Flooding
1692	04/24	New York	Severe Storms and Inland and Coastal Flooding
1691	04/20	Maine	Flooding
1690	04/02	New Mexico	Severe Storms and Tornadoes
1689	03/13	California	Severe Freeze
1688	03/14	Iowa	Severe Winter Storms
1687	03/03	Alabama	Severe Storms and Tornadoes
1686	03/03	Georgia	Severe Storms and Tornadoes
1685	02/23	Louisiana	Severe Storms and Tornadoes
1684	02/23	Pennsylvania	Severe Storms and Flooding
1683	02/22	Oregon	Severe Winter Storm and Flooding
1682	02/14	Washington	Severe Winter Storm, Landslides, and Mudslides
1681	02/09	Illinois	Severe Winter Storm
1680	02/08	Florida	Severe Storms, Tornadoes, and Flooding
1679	02/03	Florida	Severe Storms and Tornadoes
1678	02/01	Oklahoma	Severe Winter Storms
1677	02/01	Oklahoma	Severe Winter Storm
1676	01/15	Missouri	Severe Winter Storms and Flooding
1675	01/07	Kansas	Severe Winter Storm
1674	01/07	Nebraska	Severe Winter Storms

Recovery from natural disaster entails clean-up (debris removal, demolition of destroyed facilities, etc.), rebuilding of infrastructure, repairing and rebuilding homes and businesses, and addressing the needs of the poor, the elderly, and other people who are not capable of addressing their own critical issues. Natural disaster does not discriminate. Typically, natural disasters affect everybody in their path, regardless of age, race, or economic standing.

The biggest challenge in handling natural disasters is causing people to believe that disasters will happen, and that people have the ability to keep themselves from becoming victims by being better prepared. Beyond that, the next challenge is managing expectations of the public, the media, and the politicians. Then there is the challenge of seeing that the needs of all disaster victims are equitably met. Another challenge is the prevention of fraud. Insurance companies, FEMA, state and local government officials, and disaster victims are all involved in the financial issues.

Historically, the federal government has paid the larger share in disaster relief programs. Typically, the federal share of public assistance projects is 75 percent of the eligible costs, and can go to 90 percent in significant disasters. States may pay the non-federal share, or share the cost with affected local governments. North Carolina is one of the few states that have always paid the full non-federal share. Individual assistance and hazard mitigation programs are also cost-shared, with the federal share being larger than the state and local shares.

State and local government's ability to pay has probably been the most significant factor influencing allocations. North Carolina set aside $20 million for disaster response and relief in the 2006 session of the North Carolina General Assembly. Prior to the 2006 session, funding to support disaster response and relief had to be found on a case-by-case basis.

FEMA is cutting back on the administrative costs of managing disasters, which will cause states to take on more of the costs. In fact, North Carolina has chosen to have a state-appropriated staff of professionals to manage disaster relief programs. This ensures knowledgeable, reliable program managers when disasters occur in the state of North Carolina.

North Carolina also has a program for state-declared disasters that do not meet the federal threshold, but are of significant enough impact to local budgets that they require state assistance. (See www.ncleg.net/enactedlegislation/statutes/html/bysection/chapter_166a/gs_166a-6.01 .html.) States must recognize that disasters will happen, and they have a responsibility to share in the costs. Ultimately, states will have to budget for disaster response and relief programs.

When natural disasters strike, the initial expectation of individuals and groups is that their lives and situations will be returned to normal by an agency of the state or federal government. The best way for the agencies involved to manage those expectations is with two-way, face-to-face, open and honest communication. Program and project eligibility is the greatest area for misunderstanding, in my opinion. The key issue everyone must understand is that disaster relief programs have specific rules. Those affected by disaster cannot proceed on the assumption that they may take any action deemed necessary to resolve perceived problems and ask for forgiveness later. Eligibility or permission must be obtained prior to committing to specific disaster relief projects, or they may be deemed to be ineligible.

A fundamental issue in natural disaster relief is the responsibility to care for one's self, family, organization, agency, or business. As individuals, we must recognize and understand that natural disaster events will occur, and where population, development, and natural disaster intersect, there will be significant impact. It is incumbent upon all of us as participants in the natural environment, and those of us who are professional emergency managers, to recognize the hazards, plan for the hazards, advocate for building to avoid the hazards, and insure our property against the hazards. None of us can afford to assume that natural disaster will never affect us, and that if disaster does affect us, assume the government will come to our aid and return chaos to normal.

When events like Hurricane Katrina play out over the worldwide media, we tend to be guided to evaluate the federal government's response, but the questions rarely asked are the ones that pertain to what the disaster victims did to prepare and to respond for themselves. How well-prepared were the affected local governments, and what specific actions did they take to mitigate the effects, to warn and guide the citizens, to respond, and to ultimately recover from the event? What actions were taken by the affected state governments to coordinate mitigation and response actions with the local governments, to facilitate and coordinate the federal response? All disasters begin and end locally, and everybody has a stake in the outcome.

If we can accept that responsibility is a fundamental issue in natural disaster relief, then beginning with the individual and projecting through the community and all levels of government, disaster preparedness, response, and relief belong to all of us. Our safety, the protection of our property, and perhaps our very lives depend on our commitment to plan and prepare for potential disaster. A basic necessity for adequate and timely natural disaster relief is quality planning before the disaster occurs. The consequences of disaster are predictable, and based on this fact, we can better prepare ourselves, our communities, and our government at all levels, only if we will take the time to plan our actions and to coordinate those plans with each other. The responsibility is ours.

I believe there are two factors behind the issue of responsibility for natural disaster relief. First, there is the evolving nature of the business of emergency management, and second is the philosophical change in focus that occurs in our political process when we elect new leaders. Emergency management and relief from natural disaster have evolved along with the development/evolution of the Federal Emergency Management Agency (FEMA). FEMA is the implementing/managing agency for the Robert T. Stafford Disaster Relief and Emergency Assistance Act (Public Law 100-707). This act provides the statutory authority for a majority of the federal disaster recovery/relief programs.

Before the formation of FEMA under the Carter Administration at the urging of the National Governors Association, disaster response efforts were fragmented. President Carter consolidated the Federal Insurance Administration, the National Fire Prevention and Control Administration, the Federal Preparedness Agency, the Federal Disaster Assistance Administration, and the Defense Civil Preparedness Agency to form FEMA in 1979. FEMA has been shaped by its leadership and the events of our nation since it was formed. From the Three Mile Island nuclear power accident through Hurricane Hugo and the Loma Prieta earthquake,

Hurricane Andrew, Hurricane Floyd, the September 11, 2001, terrorist attacks, the 2004 hurricane season, and Hurricane Katrina in 2005, to the current events of midwestern flooding, northeastern snow storms, and an emerging drought in the southeast, the nation's preparedness program has evolved.

The success or failure of FEMA has been graphically portrayed by the media every time one of these significant disasters has occurred. Successes have tended to raise expectations and add to the false belief that FEMA can and will fix all that is wrong in the aftermath of a natural disaster. Failures, or perceived failures, have tended to cause changes in FEMA leadership, program focus, and developmental direction. The terrorist attacks of 2001 almost eliminated FEMA, and Hurricane Katrina completely changed FEMA's leadership and developmental direction for the future.

The other major factor is political philosophy. Presidents are likely going to choose a FEMA director based on the president's familiarity with, knowledge of, and experience with disaster. If the president views disaster preparedness as a priority, then he or she should seek a highly qualified professional disaster manager for the position. If the president does not view disaster preparedness as a priority, then the choice of a FEMA leader may not be that important.

Success or failure in disaster response and recovery, as viewed by the public, is greater than FEMA. It reflects on the administration, as well. The evolution of FEMA and political philosophy, combined, has developed a disaster preparedness posture in the United States where the largest share lies with the federal government. Infrequency of disaster causes preparedness to be a relatively low priority with many local and state governments. The issue of responsibility becomes one of default. If a local government chooses not to make disaster preparedness a priority, the state must come to its aid when disaster strikes to ensure that the immediate needs of the citizens (disaster victims) are met. If the state is not adequately prepared, the default is to the federal government—thus, the current disproportionate share.

Congress recognizes the critical need for emergency and disaster preparedness at all levels of government and has recommended raising the level of funding (Emergency Management Performance Grants) from $200 million in Fiscal Year 06-07 to $300 million in FY 07-08. The funding must be matched fifty-fifty at the local and state levels. It can be used to support salaries for personnel, and without it, there would be virtually no national emergency and disaster preparedness program at the local or state level.

Local and state elected leaders have choices to make, as well. Should disaster preparedness be a priority, or do they perceive disaster preparedness to be a responsibility of the state or the federal government? Remember, all disasters begin and end locally. I believe strongly that disaster preparedness must become a priority business with government at all levels. Citizens and business and industry leaders are becoming more aware of the need to prepare for disaster through the instant and constant media coverage of events as they occur anywhere in the nation, or in the world, for that matter. Business leaders recognize the need to plan for disaster and to prepare so they can return to operations as quickly as possible in the aftermath of an event. Business leaders are also actively engaged in trying to integrate their disaster response and recovery plans with those of the local governments. Efficient disaster response may be the difference between success and failure for an otherwise healthy business.

Legal Aspects

Disaster relief situations requiring legal assistance include interpretation of federal and state law, writing of legal agreements of understanding and contracts, challenging/appealing decisions, and challenging or defending actions taken because of a natural disaster. The main law associated with natural disaster relief is the Robert T. Stafford Relief and Emergency Assistance Act. (Public Law 93-288 as amended is the primary law guiding disaster relief programs.) Legal assistance is required because a natural disaster is an event out of the ordinary that causes damage, destruction, injuries, and possible loss of life. Under these circumstances, there arises the potential for disagreement and conflict. Under our system of governance, the legal system is where these conflicts may be resolved.

Every disaster relief project, or aid to a victim, requires an agreement of understanding. What happened? How will the issue be addressed and fixed? How much will it cost? Are the costs eligible under the existing programs?

Who supplies the funds? These agreements are written by, interpreted by, and challenged by attorneys. Legal skills required include the ability to review and interpret federal law, the ability to write agreements of understanding and contracts, and the ability to challenge/appeal bad decisions. Grand-scale disasters like Hurricane Katrina tend to generate a multitude of issues, many requiring immediate decisions and action. Life-and-death decisions have to be made, and questions may be raised in the aftermath, whether the decisions made were consistent with the written plans and the existing legislation. Attorneys have a significant role in these issues.

Examples of the kinds of issues that are common to disaster response include rescue operations, debris management, and restoration of critical infrastructure. Rescue operations could be the use of helicopters or swift-water search teams to rescue disaster victims in a flooding event like Hurricane Katrina. The costs of the rescue operation, to include personnel, equipment, and supplies are all potentially eligible costs under the Public Assistance Category B–Emergency and Protective Measures. Debris management is a separate Category A, devoted solely to the management and disposal of disaster-generated debris, which is one of the most costly components of disaster relief. Categories C through G deal with the restoration of critical infrastructure, like water systems, publicly owned electric distribution systems, public buildings, roads and bridges, and recreational facilities.

The following is taken directly from the Federal Emergency Management Agency's Web site, and it is a detailed description of eligible categories of work under the Public Assistance program:

> ### Categories of Work—Reference Topics
>
> There are two types of work eligible for reimbursement through a Public Assistance Grant: emergency work and permanent work. Each of these work types are further divided into categories based on the action being performed for emergency work, or the type of facility repaired for permanent work. The categories of work are often identified by a single letter. The categories are:

Emergency Work

A. Debris Removal

B. Emergency Protective Measures

Permanent Work

C. Road Systems and Bridges

D. Water Control Facilities

E. Buildings, Contents, and Equipment

F. Utilities

G. Parks, Recreational, and Other

Category A: Debris Removal

Debris Removal is the clearance, removal, and/or disposal of items such as trees, woody debris, sand, mud, silt, gravel, building components, wreckage, vehicles, and personal property.

For debris removal to be eligible, the work must be necessary to:

- Eliminate an immediate threat to lives, public health and safety
- Eliminate immediate threats of significant damage to improved public or private property
- Ensure the economic recovery of the affected community to the benefit of the community-at-large
- Mitigate the risk to life and property by removing substantially damaged structures and associated appurtenances as needed to convert property acquired through a FEMA hazard mitigation program to uses compatible with open space, recreation, or wetlands management practices

Examples of eligible debris removal activities include:

- Debris removal from a street or highway to allow the safe passage of emergency vehicles
- Debris removal from public property to eliminate health and safety hazards

- Examples of ineligible debris removal activities include:
- Removal of debris, such as tree limbs and trunks, from natural (unimproved) wilderness areas
- Removal of pre-disaster sediment from engineered channels
- Removal of debris from a natural channel unless the debris poses an immediate threat of flooding to improved property

Debris removal from private property is generally not eligible because it is the responsibility of the individual property owner. If property owners move the disaster-related debris to a public right-of-way, the local government may be reimbursed for curbside pickup and disposal for a limited time. If the debris on private business and residential property is so widespread that public health, safety, or the economic recovery of the community is threatened, FEMA may fund debris removal from private property, but it must be approved in advance by FEMA.

Category B: Emergency Protective Measures

Emergency Protective Measures are actions taken by Applicants before, during, and after a disaster to save lives, protect public health and safety, and prevent damage to improved public and private property. Emergency communications, emergency access, and emergency public transportation costs may also be eligible.

Examples of eligible emergency protective measures are:

- Warning devices (barricades, signs, and announcements)
- Search and rescue
- Security forces (police and guards)
- Construction of temporary levees
- Provision of shelters or emergency care
- Sandbagging
- Bracing/shoring damaged structures
- Provision of food, water, ice and other essential needs

- Emergency repairs
- Emergency demolition
- Removal of health and safety hazards

Category C: Roads and Bridges

Roads (paved, gravel, and dirt) are eligible for permanent repair or replacement under the Public Assistance Program, unless they are Federal-aid roads. Eligible work includes repair to surfaces, bases, shoulders, ditches, culverts, low water crossings, and other features, such as guardrails. Damage to the road must be disaster-related to be eligible for repair. In addition, repairs necessary as the result of normal deterioration, such as "alligator cracking," are not eligible because it is pre-disaster damage.

Landslides and washouts often affect roads. Earthwork in the vicinity of a road may be eligible, but only if the work is necessary to ensure the structural integrity of the road.

Road or bridge closures resulting from a disaster may increase traffic loads on nearby roads. If diverted traffic causes damage to a road, FEMA may pay to repair this damage if no alternative is available. Restoration of a damaged road may include upgrades necessary to meet current codes and standards, as defined by the State or local department of highways. Typical standards affect lane width, loading design, and construction materials.

Bridges are eligible for repair or replacement under the Public Assistance Program, unless they are on a Federal-aid road. Eligible work includes repairs to decking, guardrails, girders, pavement, abutments, piers, slope protection, and approaches. Only repairs of disaster-related damage are eligible. In some cases, FEMA may use pre-disaster bridge inspection reports to determine if damage to a bridge was present before the disaster.

Work to repair scour or erosion damage to the channel and stream banks is eligible if the repair is necessary to ensure the structural integrity of the bridge. Earthwork that is not related to the structural integrity of the bridge is not eligible. Similarly, work to remove debris, such as fallen trees, from the channel at the bridge is eligible if the debris could cause further damage to the structure or if the blockage could cause flood waters to inundate nearby homes, businesses, or other facilities.

When replacement of a damaged bridge is warranted, eligible work may include upgrades necessary to meet current standards for road and bridge construction, as defined by the State or local highway department. Typical standards affect lane width, loading design, construction materials, and hydraulic capacity. If code requires, and if the Applicant has consistently enforced that code, FEMA will permit changes in the bridge design from one lane to two lanes to include access modification for a short distance (i.e., within area of damage). This does not apply to other expansions of capacity (e.g., from two lanes to four lanes).

Permanent restoration of a road or bridge that service USACE or NRCS levees and dams, private and commercial roads, and homeowners' association roads or fall under the authority of the Federal Highway Administration is not eligible for public assistance.

Category D: Water Control Facilities

Water control facilities include dams and reservoirs, levees, lined and unlined engineered drainage channels, shore protective devices, irrigation facilities, and pumping facilities.

Restoration of the carrying capacity of engineered channels and debris basins may be eligible, but maintenance records or surveys must be produced to show the pre-disaster capacity of these facilities. The pre-disaster level of debris in the channel or basin is of particular importance to determine the amount of newly deposited disaster-related debris. Such a facility must also have had a regular clearance schedule to be considered an actively used and maintained facility.

Restoration of reservoirs to their pre-disaster capacity also may be eligible in accordance with the criteria for debris basins described above. Not all reservoirs are cleaned out on a regular basis, and evidence of pre-disaster maintenance must be provided to FEMA. In addition, removal of debris that poses an immediate threat of clogging or damaging intake or adjacent structures may be eligible.

The USACE and NRCS have primary authority for repair of flood control works, whether constructed with Federal or non-Federal funds, as well as authority over federally funded shore protective devices. Permanent repairs to these facilities are not eligible through the PA Program.

Category E: Buildings and Equipment

Buildings, including contents such as furnishings and interior systems such as electrical work, are eligible for repair or replacement under the Public Assistance Program. In addition to contents, FEMA will pay for the replacement of pre-disaster quantities of consumable supplies and inventory. FEMA will also pay for the replacement of library books and publications. Removal of mud, silt, or other accumulated debris is eligible, along with any cleaning and painting necessary to restore the building.

If an insurance policy applies to a facility, FEMA will deduct from eligible costs the amount of insurance proceeds, actual or anticipated, before providing funds for restoration of the facility. FEMA will reduce public assistance grants by the maximum amount of insurance proceeds an Applicant would receive for an insurable building located in an identified floodplain that is not covered by Federal flood insurance. The reduction in eligible costs will be the larger of the two reductions just described. The owners of insurable buildings can expedite the grant process by providing FEMA with policy and settlement information as soon as possible after a disaster occurs.

FEMA may pay for upgrades that are required by certain codes and standards. Examples include roof bracing installed following a hurricane, seismic upgrades to mitigate damage from earthquakes, and upgrades to meet standards regarding use by the disabled. For repairs, upgrades are limited to damaged elements only. If a structure must be replaced, the new facility must comply with all applicable codes and standards regardless of the level of FEMA funding.

If a damaged building must be replaced, FEMA has the authority to pay for a building with the same capacity as the original structure. However, if the standard for space per occupant has changed since the original structure was built, FEMA may pay for an increase in size to comply with that standard while maintaining the same occupant capacity. A Federal or State agency or statute must mandate the increase in space; it cannot be based only on design practices for an industry or profession.

Category F: Utilities

Typical Utilities include:

- Water treatment plants and delivery systems
- Power generation and distribution facilities, including generators, substations, and power lines
- Sewage collection systems and treatment plants
- Telecommunications

The owner of a facility is responsible for determining the extent of damage incurred. FEMA does not provide funds for random surveys to look for damage, such as TV inspection of sewer lines. If disaster-related damage is evident, however, FEMA may pay for inspections to determine the extent of the damage and method of repair.

While FEMA will pay for restoration of damaged utilities, FEMA does not provide funds for increased operating expenses resulting from a disaster. Similarly, FEMA cannot provide funds for revenue lost if a utility is shut down. However, the cost of establishing temporary, emergency services in the event of a utility shut-down may be eligible.

Category G: Parks, Recreational Facilities, and Other Items

Repair and restoration of parks, playgrounds, pools, cemeteries, and beaches. This category also is used for any work or facility that cannot be characterized adequately by Categories A-F

Eligible publicly owned facilities in this category include: playground equipment, swimming pools, bath houses, tennis courts, boat docks, piers, picnic tables, and golf courses.

Other types of facilities, such as roads, buildings and utilities, that are located in parks and recreational areas are also eligible and are subject to the eligibility criteria for Categories C, D, E, and F.

Natural features are not eligible facilities unless they are improved and maintained. This restriction applies to features located in parks and recreational areas. Specific criteria apply to beaches and to trees and ground cover.

Beaches. Emergency placement of sand on a natural or engineered beach may be eligible when necessary to protect improved property from an immediate threat. Protection may be to a 5-year storm profile or to its pre-storm profile, whichever is less.

A beach is considered eligible for permanent repair if it is an improved beach and has been routinely maintained prior to the disaster. A beach is considered to be an "improved beach" if the following criteria apply:

- the beach was constructed by the placement of sand to a designed elevation, width, grain size, and slope; and
- the beach has been maintained in accordance with a maintenance program involving the periodic re-nourishment of sand at least every 5 years.

Typically, FEMA will request the following from an applicant before approving assistance for permanent restoration of a beach:

- design documents and specifications, including analysis of grain size;
- "as-built" plans;
- documentation of regular maintenance or nourishment of the beach; and
- pre- and post-storm cross sections of the beach.

Permanent restoration of sand on natural beaches is not eligible.

Trees and Ground Cover. The replacement of trees, shrubs, and other ground cover is not eligible. This restriction applies to trees and shrubs in recreational areas, such as parks, as well as trees and shrubs associated with public facilities, such as those located in the median strips along roadways and as landscaping for public buildings. Grass and sod are eligible only when necessary to stabilize slopes and minimize sediment runoff.

This restriction does not affect removal of tree debris or the removal of trees as an emergency protective measure. FEMA will reimburse for the removal of tree debris and the removal of trees as emergency protective measures if the removal eliminates an immediate threat to lives, public health and safety, and improved property, or if removal is necessary to ensure the economic recovery of the affected community to the benefit of the community-at-large. However, FEMA will not reimburse for the replacement of these trees.

Need for Change

Our evolving way of life is driving the changes in natural disaster relief. Evolving technology, a "just-in-time" economy, the threat of terrorism, dependency on a multitude of inter-related systems and infrastructure, and a growing lack of self-sufficiency are driving these changes. Americans are ill-prepared to care for themselves in the event of a major natural disaster, thus the need for government to come to the aid of disaster victims until the electric power, the water treatment systems, the grocery stores, the gas stations, the pharmacies, the communication systems, the transportation systems, the hospitals, the restaurants, the businesses and paychecks, and the schools can be restored.

The two most significant political situations affecting natural disaster relief in modern time are the terrorist attacks of September 11, 2001, and the nation's response to the effects of Hurricane Katrina. The third political situation that will have a profound impact on the future of natural disaster relief will be the outcome of the 2008 presidential election because the newly elected president will have to know the leaders he or she will choose to manage his or her disaster relief programs. The president will have to decide whether his or her focus will be terrorism, natural disaster, or an all-hazards approach. Regardless of who the next president is, he or she will likely be tested early in his or her new role as president with the reality of a significant disaster. The response will be viewed by the world, and the evolution of the nation's disaster relief program will be set on its new path.

The most significant economic circumstances are our just-in-time economy and the rising cost of fuel. I believe the just-in-time economy provides a false sense of security. It is efficient. Product is readily available under normal circumstances, and because of the availability, people tend not to stock a reserve supply. The rising cost of fuel is causing the price of everything to rise, again causing people not to build a reserve. For years, the emergency management profession has advocated people should maintain an emergency supply of food and water sufficient to sustain them for a minimum of three days. Three days is not enough. People should be preparing for two to four weeks' worth of reserve, and rotating stock through that reserve to keep it fresh and readily available.

Outsourcing the disaster relief projects is probably the strongest trend in natural disaster relief. Everything from debris management to temporary housing, from managing massive hazard mitigation buyouts to project management at the local and state level is being done by private contractors. Natural disaster relief is a function of government. However, as disaster relief programs and projects continue to be outsourced, the number of jobs and the competitive salaries in the private sector are attracting the people who would normally fill the disaster relief jobs in government. Government program managers face a constant challenge in attracting and maintaining highly qualified and motivated personnel to manage the disaster relief programs and to manage the contractors who implement the programs.

The concerns here would have to be that the costs of disaster relief programs continue to rise, as do the expectations of the stakeholders and the disaster victims. Under the Clinton Administration, the nation's focus was on hazard mitigation and natural disaster prevention, through quality development that took into consideration the potential effects of disasters and addressed those effects. Under the Bush Administration, our focus has gone more to terrorism prevention and less to the prevention of natural disaster. Both are important, but experience indicates the more frequent occurrence and broader impact of natural disasters. There has to be a balance.

The nation's response to the terrorist attacks of 2001 and the nation's response to Hurricane Katrina are the two political factors that have had the greatest impact on natural disaster relief. Both events were documented, processed, and provided to the American public through a constant stream of media coverage. Political reaction to these two nationally significant events re-shaped the nation's disaster preparedness organization and posture, engaged the nation in war in Afghanistan and Iraq, changed the FEMA leadership, increased funding to state and local governments to build better response and recovery capabilities, improved the nation's mutual aid capabilities, and raised the nation's awareness of the potential for disaster.

The impact has been a driving sense of urgency on the part of Congress and the Department of Homeland Security to push grant dollars to local and state governments to build better response and recovery capability. The urgency to improve the nation's preparedness posture has caused short windows of opportunity to strategically plan, organize, and execute sound strategy for the development of better response and recovery capability. Many good things have been accomplished. Interoperable communications systems have improved; medical surgery capability has been expanded; search and rescue capabilities have improved; the national incident management system is being taught across the nation; mutual aid has significantly improved; and the nation's disaster response capabilities have been improved.

The improvements have been made with federal dollars. Maintenance and long-term replacement will become the responsibility of the local and state governments. In many cases, those future costs have yet to be considered, and sources of local and state funding will have to be identified.

Behind the political reaction to the terrorist attacks and Hurricane Katrina is the perception of vulnerability, government's inability to prevent the terrorist attacks, and government's less-than-expected response to Hurricane Katrina. The true corrective measures to these factors are professional leadership at all levels of government, detailed/quality planning for future events, education for all stakeholders about the response and recovery plans, and practicing the response and recovery actions necessary for future events.

Every change in natural disaster relief sets a precedent for the next disaster. Changes may be well-thought out, discussed with all the stakeholders, and agreed on before implementation, or they may be changes associated with the determination of eligibility of a project during the recovery phase of an ongoing disaster. Whatever the case, once done, all similar situations must be treated the same to be equitable.

The lack of adequate insurance is a driving factor in the actual costs of these changes. The cost of fuel is a factor. Outsourcing is a factor. Precedent set by past disasters is a factor. Poverty is a factor. Instant and continual media coverage is a factor. Politics is a factor. Savings opportunities need to be considered. Return to the hazard mitigation programs of the Clinton Administration, where our focus was on prevention of natural disaster impact on our developed communities. Reduction of hazards before the disaster makes the most sense.

I suspect the most significant changes in the area of natural disaster relief are centered on expectations. State and local politicians expect federal disaster relief programs to do more than they are designed to do. Citizens expect the government to come to their aid and return them to normal, and it cannot be done as fast, as efficiently, or as completely as they believe it should. Should we not all expect that citizens and governments alike purchase insurance for the potential of disaster?

Our exposure to disaster relief programs is fairly constant through media coverage of the nation's ongoing series of natural disasters. Our expectations of natural disaster relief are measured against our own experience, and what we perceive others have received through media coverage of events in other parts of the country. The costs of disaster relief are rising. Our expectations of disaster relief programs are rising, as well, because we are failing to acknowledge and accept our own responsibilities with regard to disaster relief. We all bear responsibility for our own readiness posture.

FEMA has changed; the Department of Homeland Security has been created; and as a result, state emergency management programs have found it necessary to adapt. For significant natural disasters, North Carolina has found it necessary to create supplemental disaster relief programs to complement the federal disaster relief programs. (See the "North Carolina Disaster Recovery Guide" at www.osbm.state.nc.us/disaster).

Since its formation in 1979 under the Carter Administration, FEMA has been the managing agency for the delivery of natural disaster relief from the federal government. With the formation of the Department of Homeland Security, terrorism preparedness and prevention became so important that natural disaster preparedness, response, and recovery ceased to be a top priority. The hurricane season of 2004 stressed the system, and Hurricane Katrina in 2005 exposed the vulnerabilities and the critical need for strong, effective leadership in the face of significant natural disasters. How would the nation fare if there were two or three major disasters at different locations in the country at the same time? Disaster preparedness and emergency management must become a priority business with all levels of government.

The struggle is one of priorities. What is perceived to be the greatest threat to America—terrorism or natural disaster? I believe the consequences of disaster are predictable regardless of cause, and I believe that natural disaster occurs far more frequently and has the potential to affect much greater geographical area than terrorism. Hurricanes are a grand example, and though it is a rarely occurring event, one can only begin to imagine the impact a naturally occurring pandemic flu would have on today's society. Preparedness throughout our nation is critical, and yet for efficiency and economics, we have built just-in-time delivery systems that depend on constant production and transportation to ensure availability of product. Disasters, natural or otherwise, tend to disrupt the just-in-time delivery systems and create shortage of supply.

Implementing Changes

Competition presents some challenges to implementing the necessary changes in disaster relief and preparation. The food, water, generators, fuel, cots, blankets, medicines, and a variety of other resources are needed at all levels of government and in neighboring jurisdictions. Many times the different stakeholders have developed agreements with the same resource vendors, creating a competition for the supply. Quality planning, as well as communication and cooperation with other stakeholders, is the only way to overcome the challenges.

I heard it said that at the beginning of the 2004 hurricane season, the expectation on the part of Florida citizens was that food, water, and supplies should be available and delivered within three days. By the end of the 2004 hurricane season, their expectations had risen to having food, water, and supplies available and delivered within three hours. People are used to unlimited availability of supply. Disasters disrupt our systems and render supply unavailable. People's needs and their expectations tend to exceed government's ability to deliver. Self-sufficiency needs to be re-introduced to our culture.

Changes now include local, state, and federal government being forced to invest in a limited stockpile of food, water, generators, and other essential equipment and supplies to be able to launch an initial response to disaster. Beyond the limited stockpiling, government has to establish contracts for the immediate purchase and delivery of goods and services that will be critical to the survival of citizens in the aftermath of an event.

Debris management has long been one of the most costly elements of disaster relief. Typically, disasters generate large volumes of debris, taxing local resources, landfills, and public works personnel. Contract labor and equipment are necessities. Typically, contracting at the time of the disaster, because of the desperate nature of the need, tends to drive the cost up. Current thinking is to establish pre-disaster debris contracts and rates so the contracts may be activated at the time of the need, and some savings may be achieved.

Significant amounts of debris generated by a disaster require large numbers of personnel and equipment for removal. Legitimate removal of the debris costs a great deal of money. Unfortunately, the debris management business offers some of the grandest opportunities for fraud because contractors have to be paid against some measure of the volume of debris removed. Debris is measured either by the cubic yard or by the ton. Either method requires very close monitoring and inspection. Efficiency and costs are driving changes in debris management. Pre-existing debris management contracts should be competitively bid, ensuring a better price and a more efficient response.

Often, changes to policy and procedure are well-thought out, discussed among stakeholders, modified for consensus, and implemented with advance notice. Other times, policy and procedure changes are made as a reaction to events of significance, implemented to solve an immediate problem and establish long-lasting precedent for the future. The administrative costs of managing disasters is changing from a sliding scale to a fixed rate, reducing the availability of federal dollars to states and locals for the management costs of administering natural disaster relief. The management of debris and contracting for debris removal and disposal is changing, adding incentives to pre-plan debris operations. Additionally, the management of the temporary housing program is changing, adding greater responsibility to state and local stakeholders for the long-term management and close-out of the program. Emergency management agencies at the state level, and city and county administrators and financial officers at the local level are the ones who are most affected by this.

Communication comes to mind as the best mechanism to coordinate improvements and changes among local, state, and federal agencies. There is frequent communication between the federal government and the states, generally facilitated through the National Emergency Management

Association (NEMA). FEMA typically seeks input from the states before the implementation of policy decisions concerning natural disaster relief programs.

FEMA will develop policy changes in draft form and seek distribution and input from the states through NEMA. States review the draft policy and make comment to FEMA through NEMA. FEMA will generally incorporate many of the state recommendations into final policy before it is published. Other opportunities to affect change exist through interaction with congressional members, and/or committees. Frequently, state emergency management directors are called on to testify before congressional committees on such things as mutual aid and disaster relief policies. The difficulties here would relate to policy changes that are made during a disaster as a response to an immediate issue that did not allow opportunity for discussion and compromise.

Impact

I believe the programs that exist are adequate for middle-income and above Americans. The most significant problems encountered are those that affect low-income disaster victims. Typically, the low-income population is struggling to get by on a daily basis. Disaster programs are designed, at best, only to get the disaster victim back close to where he or she was before the event. For low-income individuals and families, the programs leave a great deal of need that must be met by volunteer organizations or by supplemental state programs. Perhaps government should focus on addressing poverty with education and opportunity to work and earn an adequate living, as a way of helping prepare for natural disasters. Either we address our poverty issues and seek to build a more self-sufficient America at all economic levels, or the cost of disaster relief will continue to rise.

Controversies or challenges to preparedness lie in our increasing dependence on advancing technology, reliance upon a just-in-time economy, and lack of self-sufficiency. This renders us all to the brink of disaster when natural events and populations intersect. Improved preparedness is a necessity.

The challenges revolve around helping disaster victims understand that there is a disaster relief process, educating them on the actions they must take to be able to take full advantage of the programs that are available, and attempting to guide them toward good decisions that will improve their disaster circumstance.

Changes to the natural disaster relief programs tend to make the process of natural disaster relief difficult, complex, and not necessarily disaster-victim friendly. Going through a natural disaster with the damage or destruction of your worldly possessions and injuries or deaths of friends or family members produces significant trauma. Many disaster victims need to be counseled or guided through the recovery and relief process. FEMA's process is geared to tele-registration and going through the system by telephone. Many disaster victims need more guidance than is made available.

Disaster assistance available to disaster victims, thoughtfully applied, can make a significant difference in a victim's recovery. Borrowing money from the Small Business Administration is a part of the assistance that is available, and yet many victims allow the fact that the money is a loan and has to be paid back to be a deal breaker. They stop the process because they do not want to borrow any money. Ultimately, citizens are served or not served by the choices they make regarding their investment of the disaster relief they receive. Without sound guidance, many disaster victims squander their disaster relief dollars and fail to improve their situation.

Generally, I believe affected groups, organizations, and agencies are pleased to learn there are programs available to aid in their recovery from natural disaster. I also believe that knowledge is accompanied with a level of frustration when it comes to addressing the details of the disaster relief programs. In North Carolina, we have made it a practice to conduct daily conference calls with the affected communities that choose to participate, immediately following a disaster, to address their questions and help them better understand the process. This is not something they do routinely. Primarily, my contact is with city and county managers. If we had the manpower, a dialogue with the disaster victims would be of value.

I believe it is our responsibility to attempt to educate those who need to access the disaster relief programs on the extent and limits to the programs, and on the process for achieving maximum benefit. If potential disaster victims better understood the extent and the limits of the disaster relief programs, they might be prompted to take more responsibility for their own preparedness.

To anyone trying to navigate the natural disaster relief changes, I recommend first they go to the Federal Emergency Management Agency's Web site at fema.gov, and research disaster recovery programs. There is a wealth of information available at that site to thoroughly explain the natural disaster relief programs. I believe there has to be a constant dialogue between the states and the federal government concerning the disaster relief programs. States must strive to stay current with federal policy and procedure, but the only way to do that is to maintain dedicated full-time staff to monitor the programs and potential changes to the programs.

The Internet provides ready access to current policy and procedure, but full-time dedicated personnel are critical to keeping the state current on federal programs. The state has a responsibility to educate the local governments, whether before the event or during the event, but most local governments will not be current on federal disaster relief programs.

The only obstacle to success is the fear of failure. Success lies in persistence, open and honest two-way communication, research, quality listening, and a desire to learn and understand.

Everybody has a share in the responsibility of preparing for natural disaster. If you believe the government will take care of your needs for you, you will likely be sorely disappointed at some point. Natural disasters will happen. You can be ready or not. If you are ready, you are much less likely to become a victim.

Someday, more than one significant disaster will occur at the same time. If you have not experienced the effects of a significant disaster, likely you will some time in your life. You have a choice whether to become a disaster victim, or be prepared.

E. Douglas Hoell Jr. received a bachelor's degree in recreation and parks administration in 1975 from North Carolina State University. The following year, he began his career in emergency management as an administrative officer for Raleigh–Wake County Emergency Preparedness. He has worked for FEMA in the Radiological Emergency Preparedness program; he has worked as a contractor through Valdosta State College delivering FEMA career development training; and he has worked for the state of North Carolina, Division of Emergency Management, since 1985.

With thirty years of emergency management experience, Mr. Hoell was offered the opportunity to serve Governor Easley and the citizens of North Carolina as director of the North Carolina Division of Emergency Management, and was sworn in on July 1, 2005.

Mr. Hoell also serves as chairman of the Emergency Management Assistance Compact (EMAC) for the National Emergency Management Association (NEMA). He is a member of the NEMA board of directors. He serves as vice chairman of the North Carolina State Emergency Response Commission, and in October 2006, Governor Easley appointed Mr. Hoell to co-chair the Governor's Hazardous Materials Task Force.

Managing Expectations in Disaster Relief through Communication and Preparation

Susan Reinertson

Emergency Management

People

Many groups and individuals outside of emergency management expect and assume that declarations are declared simply. There is a process that needs to be followed, and program limitations when the process results in a declaration. They expect to be made whole after a disaster (by the federal government). The biggest challenge I see is that people get complacent and get caught by surprise when a disaster occurs, when they could have taken steps to prepare themselves to be ready for an event to occur.

While events will occur and no preparation can be done, there are things everybody can do to lessen the effects of a disaster. My role and other leaders' roles in disaster relief is to help lead change. I direct the coordination and facilitation of all the resources that agencies at all levels of governments bring to disasters. We are the meta-leaders, which is that interagency, intergovernmental coordination I just discussed. Successful disaster relief cannot be accomplished without effective leadership at all levels, which includes not only the leaders at the top, but every other employee who has a responsibility to carry out the vision and mission of their respective agencies.

The meta-leadership is leading vertically and horizontally, meaning across agencies and people, at all levels of government. I believe all employees can be effective leaders if they feel empowered to do their jobs and are given the tools to accomplish it. I think this is the best way to manage expectations of citizens, elected officials, media, and others, through the National Response Framework that was released on January 22, 2008. The framework is a national guide that addresses all jurisdictions and is built on engaged partnerships (meta-leadership). It is this partnership and leadership that will make the framework successful by educating and managing expectations.

The agencies attempt to meet these needs and expectations most effectively by extensive outreach and building relationships before disaster occurs. Improving partnerships at the state and local levels is imperative to disaster relief efforts, both before and after a disaster, as well as establishing a network of communication in each community to aid in distributing relief and determining the areas of need. It is all part of the planning effort as a whole (local, state, federal, private, etc.) to fill in the gaps to determine areas of need and how to accommodate. It is so important to be aware of the shortfalls ahead of disaster so resources can be pre-identified, as well as pre-identifying the gaps so work can be done to find out who can fill the gaps. Preparing effectively includes planning, organizing, equipping, training, exercising, and evaluating performance on a regular basis.

The biggest misunderstanding, which hampers effectiveness, is that the federal government wants to take over for local or state governments. For citizens, it is a misunderstanding that the federal government will make victims whole. It is key to understand that we all have a personal responsibility to prepare for and respond to disasters. The Federal Emergency Management Agency, or FEMA, wants to partner with local and state governments, not take over. Homeowners and flood insurance policies are essential, as well.

The National Incident Management System provides a systematic, proactive approach to guiding departments and agencies at all levels of government, the private sector, and nongovernmental organizations to work seamlessly to prepare for, prevent, respond to, recover from, and mitigate the effects of incidents, regardless of cause, size, location, or complexity, to reduce the loss of life and property and harm to the environment (from the National Incident Management System, August 2007). Those who may be involved include, but are not limited to, elected officials, such as governors and mayors; the department of transportation at city, county, state, and federal levels; the state water commission; the attorney general's office; health and human services and emergency management offices; Homeland Security; local and state law enforcement; forestry departments; and the FBI.

Speaking to a federal response, overall coordination of federal incident management activities is the responsibility of the Department of Homeland Security (DHS). The federal government maintains a wide array of capabilities and resources that can be made available to assist state, tribal, local, or other federal partners. There are various roles, depending on the organization. Specific guides are available to local, state, federal, private-sector, and nongovernmental agencies and citizens on the National Response Framework Web site. Emergency Support Functions (ESFs) group government and certain private-sector capabilities into a structure to provide the support, resources, program implementations, and services most likely to be needed during a response. There are fifteen of these groups, and they serve as the primary operational-level mechanism to provide assistance:

1. *Transportation*: Coordinated by the Department of Transportation; aviation/airspace management and control; transportation safety; restoration and recovery of transportation infrastructure; movement restrictions; damage and impact assessments
2. *Communications*: Coordinated by DHS-National Communications System; coordination with telecommunications and information technology industries; restoration and repair of telecommunications infrastructure; protection, restoration, and sustainment of national cyber and information technology resources; oversight of communications within the federal incident management and response structures
3. *Public Works and Engineering*: Coordinated by Department of Defense-Army Corps of Engineers; infrastructure protection and emergency repair; infrastructure restoration; engineering services and construction management; emergency contracting support for life-saving and life-sustaining services
4. *Firefighting*: Coordinated by Department of Agriculture-U.S. Forest Service; coordination of federal firefighting activities; support to wildland, rural, and urban firefighting operations
5. *Emergency Management*: Coordinated by DHS-FEMA; coordination of incident management and response efforts; financial management; incident action planning
6. *Mass Care, Emergency Assistance, Housing, Human Services*: Coordinated by DHS-FEMA; mass care; emergency assistance; disaster housing; human services
7. *Logistics Management and Resource Support*: Coordinated by General Services Administration and FEMA; resource support in the form of facility space, office equipment and supplies; comprehensive national incident logistics planning, management and sustainment capability
8. *Public Health and Medical Services*: Coordinated by Department of Health and Human Services; public health, medical, and mental health services, mass fatality management

9. *Search and Rescue*: Coordinated by DHS-FEMA; life-saving assistance; search and rescue operations
10. *Oil and Hazardous Materials Response*: Coordinated by Environmental Protection Agency; oil and hazardous materials (chemical, biological, radiological, etc.) response; environmental short- and long-term cleanup
11. *Agriculture and Natural Resources*: Coordinated by Department of Agriculture; nutrition assistance; animal and plant disease and pest response; food safety and security; natural and cultural resources and historic properties protection; safety and well-being of household pets
12. *Energy*: Coordinated by Department of Energy; energy infrastructure assessment, repair, and restoration; energy industry utilities coordination; energy forecast
13. *Public Safety and Security*: Coordinated by Department of Justice; facility and resource security; security planning and technical resource assistance; public safety and security support; support to access, traffic and crowd control
14. *Long-term Community Recovery*: Coordinated by DHS-FEMA; social and economic community impact assessment; long-term community recovery assistance to states, tribes, local governments, and private sector; analysis and review of mitigation program implementation
15. *External Affairs*: Coordinated by DHS; emergency public information and protective action guidance; media and community relations congressional and international affairs; tribal and insular affairs

When ESFs are activated, they may have a headquarters and regional and field presence. Each ESF comprises a coordinator and primary and support agencies. I listed the coordinator in the list above. Support agencies are activated depending on resources and capabilities for a given area.

FEMA and its partners successfully responded to more than 1,300 major disasters and emergencies across the nation since its inception in 1979. This decade has seen a number of pivotal events that presented new challenges for the nation's all-hazard emergency management system, including the September 11 terrorist attacks, a rapid succession of strong 2004 hurricanes that affected Florida, and a record 2005 hurricane season that resulted in devastating Gulf Coast impacts from Hurricanes Katrina and Rita. President Bush issued a major disaster declaration for the California wildfires and ordered greater federal aid to supplement state and local response activities in the affected areas. Federal resources began mobilizing immediately and authorized federal funds to reimburse the state for certain costs incurred under FEMA's Fire Management Assistance Grant Program. There have also been several major disaster declarations and emergencies declared for severe storms and flooding.

A hugely successful gap analysis began in 2007 in the hurricane-prone states. After refining, it will expand to all regions for all hazards in all jurisdictions. It is a tool to enable a collaborative partnership with all stakeholders to identify needs and vulnerabilities. The goal is to enhance the overall preparedness in all communities.

Financial Components of Disaster Relief

Guidelines are established to ensure disaster relief funds are used most efficiently and with as little waste as possible to meet the challenges of making sure the money goes where needed, as quickly as possible.

There is state-to-state assistance through interstate mutual aid agreements, such as the Emergency Management Assistance Compact (EMAC). It is a congressionally ratified organization that provides structure to the process. It is through this process that states can request and receive assistance from other member states. When an incident overwhelms or is anticipated to overwhelm a state, the governor can request federal assistance in the form of funding, resources, and critical services. The federal government supports the local, tribal, and state governments.

When state capabilities are exceeded, the governor can request federal assistance, including the Robert T. Stafford Disaster Relief and Emergency Assistance Act. The Act authorizes the president to provide financial and other assistance to state and local governments and certain private non-profit organizations and individuals to support response, recovery, and mitigation efforts following emergency or disaster declarations.

If an emergency is declared, funding can come from the president's Disaster Relief Fund, which is managed by FEMA, and the programs of other Federal departments and agencies. Some states have similar or match federal recovery programs. FEMA partners with state, tribal, and local jurisdictions to plan, train, and exercise. States receive Emergency Management Program Grants to support emergency management agencies. There are also homeland security grants given to local, tribal, and state jurisdictions.

Laws

Some disaster relief situations require legal assistance; they are generally handled by headquarters. I work in the regional offices, so I don't often hear about all the legal assistance being provided. Having said that, I can say eligibility factors often need a legal review based on environmental or historic issues related to a project, or there may be eligibility issues related to individuals who have specific needs that aren't traditional to a program.

There are many laws associated with natural disaster relief. I have listed some below, but according to the NRF, the authorities are numerous. They can be found at http://www.fema.gov/pdf/emergency/nrf/nrf-authorities.pdf.

The Homeland Security Act of 2002, Pub. L. No. 107-296, 116 Stat. 2135 (2002) (codified predominantly at 6 U.S.C. §§ 101-557), as amended with respect to the organization and mission of the Federal Emergency Management Agency in the Department of Homeland Security Appropriations Act of 2007, Pub. L. No. 109-295, 120 Stat. 1355 (2006), established a Department of Homeland Security as an executive department of the United States.

The Robert T. Stafford Disaster Relief and Emergency Assistance Act, Pub. L. No. 93-288, 88 Stat. 143 (1974), codified in 42 U.S.C. §§ 5121-5206 (2005), was also amended in the Department of Homeland Security Appropriations Act of 2007, Pub. L. No. 109-295, 120 Stat. 1355 (2006), particularly Title VI, the Post-Katrina Emergency Management Reform Act of 2006.

The Post-Katrina Emergency Management Reform Act (PKEMRA), which is Title VI of the Department of Homeland Security Appropriations Act, 2007, Pub. L. 109-295, 120 Stat. 1355 (2006), clarified and modified the Homeland Security Act with respect to the organizational structure, authorities, and responsibilities of FEMA and the FEMA Administrator.

Changes

For FEMA, the largest current issue in natural disaster relief is getting past the scrutiny of Hurricane Katrina. The Post-Katrina Emergency Management Reform Act, passed by Congress and signed by the president in October 2006, sets forth an expanded mission for FEMA. This act established new leadership positions within DHS, brought preparedness functions into FEMA, created and reallocated functions to other components within the department, and amended the Homeland Security Act. Our mandate is to reduce the loss of life and property and protect the nation from all hazards, including natural disasters, acts of terrorism, and other man-made disasters, by leading and supporting the nation in a risk-based, comprehensive emergency-management system of preparedness, protection, response, recovery, and mitigation. Our challenge is to achieve our vision and fully execute this mission to create a safer and more secure America.

It is everyone's responsibility (tribal, local, territory, state, and federal governments; private non-profits; private sector; and citizens) to prepare for and respond to emergencies. Rather than wait until assistance is requested, FEMA partners with local and state partners to provide assistance before their own systems are overwhelmed. FEMA does not take over, but rather partners with local authorities and organizations, being there to help if resources are needed.

The National Strategy for Homeland Security guides, organizes, and unifies the nation's homeland security efforts. It provides a framework by which the nation should be focused not only on preventing and disrupting terrorist attacks, protecting its people and critical infrastructure/key resources, and responding to and recovering from incidents that do occur, but also continuing to strengthen the foundation to ensure long-term success by creating and transforming homeland security principles, systems, structures, and institutions. According to the National Response Framework, we can do this by building a culture of preparedness, improving incident management, better utilizing science and technology, and leveraging all instruments of national power and influence. For instance, rather than wait until disaster strikes, the culture of preparedness can be improved by focusing efforts on coordinated planning and spending grant money to improve structures so that they are less affected by a disaster.

Government systems are changing business processes to respond to current issues in natural relief disaster, and FEMA is greatly affected. The changes affect all the agencies we do business with (i.e., the ESFs outlined above). Each agency acts under its own authorities and laws, but we need to do what we can to respond quickly, efficiently, and effectively.

Government programs are made available to communities through partner organizations (usually state, local, and tribal) and are intended to supplement the response activities and recovery programs of states. Most of our assistance programs offered through FEMA are authorized under the Stafford Act.

FEMA's programs are designed to assist states and communities in carrying out their responsibilities and priorities. Our assistance is available in varying forms, such as grants, technical assistance, and planning assistance to address the impacts of disasters and to take steps to reduce the potential impacts. Assistance that is made available to states, tribes, communities, and individuals following disasters includes:

- The Public Assistance Program, which provides assistance for the restoration of public and certain private nonprofit facilities damaged by an event, and the reimbursement of the costs associated with emergency protective measures and debris removal
- The Individuals and Households Program, which helps ensure that the essential needs of individuals and families are met after disasters so that they can begin the road to successful recovery
- The Hazard Mitigation Grant Program

The Pre-Disaster Mitigation Grant Program, authorized under the Stafford Act, and the Flood Mitigation Assistance Program, authorized under the National Flood Insurance Reform Act of 1994, as amended, are pre-disaster programs. The National Flood Insurance Program (NFIP), authorized by the National Flood Insurance Act of 1968, as amended, is available also.

The Pre-Disaster Mitigation grant program was authorized by Congress under the Disaster Mitigation Act of 2000, which was signed into law on October 30, 2000. This program is available to communities through the state emergency management organizations and is designed to fund nationally competitive mitigation projects and planning efforts of states and communities, as identified and prioritized in state and local mitigation plans. The development and adoption of these state and local mitigation plans are required under the Stafford Act because of the legislative amendments of 2000. Funding for this competitive grant program is not triggered by a presidential disaster declaration; rather it is funded through the annual appropriations process. All states and communities throughout the nation that have FEMA-approved mitigation plans and are enrolled in the National Flood Insurance Program are eligible to apply for the program. Accordingly, the Pre-Disaster Mitigation Grant Program will help sustain an enhanced national mitigation effort year-to-year, as opposed to previous years, when FEMA mitigation assistance was generally available only after a disaster was declared in a state.

Examples of projects funded under the program include the development of all-hazard mitigation plans, seismic retrofitting of critical public buildings, and acquisition or relocation of flood-prone properties located in the floodplain, just to name a few. All projects submitted are developed at the state or local level, must be cost-effective and technically feasible, and are approved following a nationally competitive peer-review process. In fiscal year 2007, the funding level was $100 million nationwide. Specific project applications are capped at a federal share of $3 million per project.

The Hazard Mitigation Grant Program (HMGP) is available to states and communities following presidential disaster declarations. This program has requirements similar to the Pre-Disaster Mitigation Grant Program previously described, though funds become available only after a disaster is declared and are available anywhere within the state in which the declaration was made. The amount of assistance available under the Hazard Mitigation Grant Program is a percentage of FEMA's assistance made available under the response and recovery programs.

As with the Pre-Disaster Mitigation Grant Program, all projects are developed at the state or local level, must pass a benefit-cost analysis, and are recommended by the state in accordance with the State Hazard Mitigation Plan. Again, examples of projects eligible for HMGP and the PDM grant funds include the development of all-hazards mitigation plans at both state and local levels, the seismic retrofitting of critical public buildings, and acquisition, relocation, or elevation of flood-prone properties located in the floodplain. FEMA does not fund major flood control projects or provide assistance for activities for which another federal program has a more specific or primary authority to provide.

FEMA's Flood Mitigation Assistance Program is authorized for mitigating structures insured by the NFIP within a community participating in the NFIP. Projects include the elevation, relocation, and acquisition of flood-prone structures. Because this program is funded by monies collected from NFIP policyholders, the recent focus of the program has been on mitigating repetitive loss structures to reduce the drain on the National Flood Insurance Fund. Severe repetitive loss properties are defined as properties that have experienced four or more flood losses of at least $5,000 each, with at least two claims payments occurring in a ten-year period, and with the total claims paid exceeding $20,000; or properties that have received at least two separate flood claims payments, where the cumulative flood claims payments exceed the value of the property.

There are significant eligibility and funding challenges to FEMA and its state partners in developing successful mitigation projects, including relocation. With respect to eligibility, projects that receive FEMA grant funding must demonstrate a positive benefit-cost ratio. The benefit-cost requirement for all federal grants that FEMA is required to apply to its grants programs is outlined in OMB Circular A-94. Basically, an applicant must demonstrate that the benefit of the project is the same as or greater than the cost. Without a positive benefit-cost ratio, a project is not eligible for funding consideration.

The Pre-Disaster Mitigation Grant Program had a total appropriation of $100 million for Fiscal Year 2007, with a $3 million cap on each nationally selected project and a limitation that no one state might receive more than 15 percent of the total available funding. Additionally, the non-federal cost-share requirements of our mitigation grants can pose a problem if a state passes on these costs to local communities because of limited financial resources. By statute, the federal government will pay 75 percent of the eligible costs, with the remaining 25 percent paid by either the state or the local government.

The Post-Katrina Reform Act empowered the regions. While I work on implementing Administrator R. David Paulison's vision, I do that through creating the culture of preparedness for the entire region (Region X), which includes all the states, not stopping at the borders. I believe in integrated preparedness. While each state has its own uniqueness, vulnerabilities, and requirements, the region can build on that by integrating what keeps them different and what they share. I would like to see a mapping of regional preparedness. We will be able to get there through a gap analysis.

FEMA's Administrator Paulison is committed to building the nation's preeminent emergency management and preparedness agency. He is doing this by building around nine core competencies: incident management, operational planning, disaster logistics, emergency communications, service to disaster victims, continuity programs, public disaster communications, integrated preparedness, and hazard mitigation.

The competencies apply to all, but there are leads for all, just as in the ESFs. We have built stronger leadership, teamwork, and accountability throughout the nation; the agency has improved the professional workforce by supplying information, support, and resources that employees need to do their jobs; stronger partnerships have been built, which have resulted in stronger coordination among FEMA, Department of Homeland Security, and stakeholders external to FEMA and DHS.

The improvement for all is that FEMA is strengthening the nation's ability to address disasters, providing consistent and reliable information during peace time and disaster time, investing in the workforce, and changing the business culture to reward performance, build public trust, and instill confidence in the department.

On February 28, 2003, the president issued Homeland Security Presidential Directive–5 (HSPD–5), *Management of Domestic Incidents*, which directed the Secretary of Homeland Security to develop and administer a National Incident Management System (NIMS). This system provides a consistent nationwide template to enable federal, state, tribal, and local governments, the private sector, and nongovernmental organizations to work together to prepare for, prevent, respond to, recover from, and mitigate the effects of incidents regardless of cause, size, location, or complexity. This consistency provides the foundation for utilization of NIMS for all incidents, ranging from daily occurrences to incidents requiring a coordinated federal response (www.fema.gov/pdf/emergency/nrf/nrf-nims.pdf).

A collaborative process, integrating local, state, tribal, territory, and federal government and private-sector and nongovernmental organizations, is being used to coordinate improvements and changes. This is done through work groups, written input, and conference calls.

Region X's Regional Interagency Steering Committee works directly with federal Emergency Support Function leadership and discusses state-to-state support. It allows us to speak to updates of federal plans to support state operations. It also allows the FEMA region and other federal agencies in the region to hear directly from the states to understand state-specific initiatives and priorities.

The Post-Katrina Reform Act also mandated each region to form a Regional Advisory Council. The councils are made up of elected officials, emergency managers, and response providers from state, local, and tribal governments. Region X's council held its first meeting in late November in Bellevue, Washington. A primary goal of the council will be to improve communication and understanding among the various organizations involved in emergency management and response.

For anyone trying to navigate the changes now occurring, and occurring in the future in natural disaster relief, the optimum approach is to get involved, stay involved, and learn what everyone brings to the table, what limits are placed on agencies. Pay attention to the work being done on the National Response Framework and its accompanying appendices. The best resources for this are the Web sites *www.fema.gov and www.dhs.gov;* national associations, including the National Emergency Management Agency, International Emergency Management Association, National Governor's Association, U.S. Department of Homeland Security, Federal Emergency Management Agency, as these guide users to local, state, and tribal contacts; press releases; and state and local homeland security and emergency management organizations.

Susan Reinertson was appointed regional administrator of Region X for the Department of Homeland Security's Federal Emergency Management Agency (FEMA) in October 2006. She is responsible for coordinating FEMA mitigation, preparedness, and disaster response and recovery activities in four states in the northwest—Alaska, Idaho, Oregon, and Washington. She has experience in all aspects of emergency management and homeland security programs, communications, and the legislative process at the local, tribal, state, and federal level.

Ms. Reinertson comes to FEMA from the Homeland Security Institute in Arlington, Virginia, where, as an analyst, she coordinated project development of a system integrating local, state, and territorial governments and the federal government to assess national preparedness.

From 1997 to early 2006, Ms. Reinertson worked for the State of North Dakota—Office of the Adjutant General. Her most recent position there was director of Homeland Security and Emergency Management. She directed a $57 million budget and was responsible for the statewide homeland security and emergency management strategic plan. Earlier, as deputy director, she coordinated, implemented, maintained, and evaluated statewide homeland security, communications, and emergency management programs and handled budgetary responsibilities. Her experience also includes positions as operations and planning officer, public assistance officer, alternate state coordinating officer, and governor's authorized representative, as well as emergency management positions for local city/county government.

Ms. Reinertson received both a Bachelor of Arts degree in speech-language pathology and a Master of Public Administration degree from the University of North Dakota. She earned a Master of Arts degree in security studies (homeland security and defense) from the Naval Postgraduate School in California, and will complete the National Preparedness Leadership Initiative at the John F. Kennedy School of Business at Harvard University in June 2008.

Acknowledgement: *I want to acknowledge Dennis Hunsinger, deputy regional administrator for FEMA, Region X, for his contribution to this chapter.*

The Challenges Faced When Recovering from a Catastrophic Event

Mike Womack

Executive Director

Mississippi Emergency Management Agency

Providing Relief for Citizens and Communities Affected by Catastrophic Disasters

The common misconception is that the federal government will be able to cover everyone's losses following a major disaster. Hurricane Katrina was the worst natural disaster our country had ever experienced. Thousands of communities and households across multiple states were obliterated. The federal government would have to provide assistance, but for many individuals this assistance alone would not be enough.

The first federal funds made available for families affected by the hurricane were from the Federal Emergency Management Agency through its Individual Assistance Program. This assistance was for those individuals whose homes were uninsured or underinsured at the time of Hurricane Katrina. The maximum amount that FEMA could provide a household in 2005 was $26,200. Obviously, that alone would not be enough to rebuild someone's home.

The funding shortfall to help individuals affected by the storm was obvious in the days, weeks, and months that followed August 29, 2005, and more programs were made available to assist Mississippi residents affected by the storm. Those programs included the thousands of non-profit groups who arrived from all over the world with supplies and a volunteer workforce to help rebuild the homes of people they never knew. There were also federal programs, such as the U.S. Small Business Administration disaster loan program, which provides low-interest disaster loans to residents and businesses, and the Homeowner's Grant Program funded by Congress.

The application and approval period for these federally funded programs, though, can be lengthy and frustrating for everyone involved. The necessary oversight and paperwork involved to make certain that only those people who are really deserving of the assistance receive the assistance takes time. It can be discouraging to those who need to rebuild or repair their homes immediately so that they can move from temporary housing into their homes again.

Assisting Local Governments

In a small-scale disaster, local governments could make repairs or rebuild damaged public structures using their own funds while continuing to take in tax revenue from local businesses and residents. For a large-scale disaster such as Katrina, though, people do not realize that when the local businesses, homes, and infrastructure were destroyed, so was the tax base the local government uses to survive and help fund the rebuilding process.

To bring back their local tax bases, communities need infrastructure, like sewer and water, as well as critical services, such as hospitals and law enforcement. After Hurricane Katrina, local governments knew that these projects could not be funded without the help of the federal government. The federal government funds such projects through FEMA's public assistance program. More than $2.3 billion has been allocated to help fund public assistance projects, but there is also a cost share associated with that, and in the Stafford Act, the law that governs emergency management, the federal government will pay 75 percent for these projects, and the local and state government will split the remaining 25 percent. In Mississippi, the state pays 12.5 percent, and the local government will pay 12.5 percent. While 12.5 percent does not seem like very much, for many communities, 12.5 percent of their cost share for public assistance projects would have been millions of dollars. This is money that these communities did not have. Thankfully, though, Congress saw that this catastrophic event was going to require that some of the rules be waived and directed that all public assistance projects be 100 percent federally funded for this disaster.

The federal government also fully funded FEMA's Other Needs Assistance Program. ONA falls under the Individual Assistance Program and funds costs such as vehicle replacement for someone who lost a vehicle in the disaster, and it can pay medical and transportation expenses that occurred as a result of the disaster.

A large-scale disaster such as Katrina requires assistance from so many different local, state, and federal agencies, from the Mississippi Department of Transportation, which worked tirelessly with its federal partners to rebuild bridges to reconnect devastated communities, to the Mississippi Department of Human Services, which worked with its federal partners to provide disaster food stamps and disaster unemployment to those affected by the storm, and the Mississippi Department of Finance and Administration, which helps disburse billions of dollars in relief to local governments. We all work together to recover from this disaster, with the tremendous support of our governor, Haley Barbour, and his staff.

How Attorneys Can Play a Role in Disaster Recovery

Those affected by disasters require a variety of legal assistance, and I do not know that we really saw the wide range of roles that lawyers served in a disaster until Hurricane Katrina.

In my opinion, housing and rebuilding were the areas that legal services were most prevalent. When someone's house deed, insurance documents, and other important documents are lost in a disaster, it can be difficult to prove who had legal ownership of the home.

Many people have sought legal council to assist them with their insurance claims, as well. Homeowners who did not live in a moderate- or high-risk flood zone may not have carried flood insurance on their property, but carried only a wind policy. When they approached their wind policy carrier for repayment, many insurance companies told their clients that it was the water and not the wind that destroyed their home. Thousands of homeowners who found themselves in this position took their insurance provider to court to let a judge decide whether their insurance provider should pay or not.

The state also uses lawyers to help manage the appeals process for the more than $2.3 billion in public assistance funds. Sometimes local governments will spend funds they were given for a project, only to find that FEMA or the Office of Inspector General will later require the funds be returned. At that point, the local jurisdiction can submit an appeal, and they might see legal assistance to help them with that appeal.

I have worked with lawyers on a variety of emergency management issues, and I find that it is most beneficial when they have an undergraduate degree in the field in which they are working now. For instance, a lawyer with an accounting degree can help tremendously if FEMA is trying to take funds away from a public assistance applicant because the bookkeeping was insufficient. A lawyer with accounting experience would be able to see what exactly FEMA meant and what the local government needed to do to correct this and approve the funding.

Rebuilding Housing

The extension of the temporary housing program, which provides free housing or rental assistance, has been a tremendous help for citizens affected by Hurricane Katrina, but it also serves as constant reminder of the lack of affordable housing available.

There are many challenges homeowners and renters face when rebuilding their homes or finding affordable housing after a catastrophic disaster such as Katrina. Coastal communities are traditionally very valuable, and the closer you get to the water, the more valuable your land becomes. On the Mississippi Gulf Coast before the hurricane, there were homes along the waterfront that ranged in value from $500,000 to $1 million. Further inland were homes that were more modest, cottage-style homes built more than sixty years ago for $20,000 to $30,000, but would now cost $150,000 or more to rebuild. Someone who lived there and was uninsured may not be able to afford that.

Because so many communities were left with nothing but slabs, many local building officials are taking this opportunity while everyone rebuilds to tighten zoning regulations. This has significantly affected our alternative housing program. Our alternative housing program was the result of a $280 million federal grant to design a safer, more attractive alternative to a travel trailer that FEMA currently uses for disaster housing. For the program, we designed one-, two-, and three-bedroom cottages that fit in perfectly with architecture along the Gulf Coast. These cottages are available to all eligible residents of FEMA temporary housing on the Mississippi Gulf Coast, but only if their jurisdiction will allow the units to be placed there. The problem that we have run into is that many jurisdictions and citizens feel these temporary housing units will become permanent and cause their property values to go down.

The public housing stock in the communities was destroyed, and there is a reluctance of local communities to rebuild a large amount of public housing. Also, renters are seeing rental prices skyrocket. An apartment that used to rent to a low- to middle-income family for $700 month may now be leased to a middle- to upper-income family that is rebuilding their home for $1,200 a month. This has been the governor's main focus—to rebuild affordable housing stock. He has developed many programs and incentives for developers to rebuild affordable housing, but the marketplace is really driving the rebuilding process. Most of the housing that has been rebuilt is middle- to upper-income housing, so the working class has few places to live, so the workforce needed to rebuild the Mississippi Gulf Coast is in extremely short supply.

Changing the Way We Govern Disasters

I don't think that the Stafford Act was written with a catastrophic event in mind or that we could ever have imagined something as catastrophic as Hurricane Katrina. Therefore, some changes need to be made in how we govern disasters.

In the area of individual and household assistance, $26,200 was not enough to rebuild someone's home, and the new threshold of $28,200 will still not be enough. We were lucky that Congress allocated the additional funds that we used to provide homeowners with grants to rebuild and repair, but there is no law that guarantees that they will do that for every catastrophic event.

There also needs to be more done to help replace local tax revenues. The current law does not allow FEMA to assist local governments with working capital. The cities and counties in a catastrophic event will run into the same problems that the cities and counties did after Hurricane Katrina—the risk of not being able to pay city and county workers and losing vital workforce that you need to have in place to bring residents back, such as county hospital employees and teachers.

Also, the appeal process and taking money back from communities that are devastated needs to be reevaluated. FEMA is bound by rules and regulations, so I do not blame it for how it administers the program. With all the rules and regulations and dealing with various federal offices, it can literally be a nightmare for local governments to keep it all straight. I would like to see major reform in the entire process and a focus on the oversight of those who have cheated the government.

With many states involved, it is also important that the same federal policy one state has to follow also be followed by other states. That is why the federal office of Gulf Coast Recovery was established—to ensure that the federal policies are the same for all states affected by Hurricanes Katrina and Rita.

Another important policy change for this disaster involved parochial schools. Many of the schools along the Mississippi Gulf Coast are parochial schools, meaning they are faith-based. The majority of those schools are Catholic schools. The public assistance program does not usually fund the rebuilding or repairing of these schools, but because 20 percent to 30 percent of the school-age children on the Mississippi Gulf Coast attend those schools, not rebuilding them would put a tremendous burden on the public school system. It was decided that for this disaster, the federal government would fund the rebuilding and repairing of these schools.

Citizens and local governments have already benefited from many policy changes during this disaster, such as debris removal extensions, the elimination of the cost share, and the extension of temporary housing.

In the future, though, as we see more and more disasters like Katrina, if we do not make those policy changes now, we will be faced with the same challenges again.

Preparing for the Next Catastrophic Event

The recovery from Hurricane Katrina will take many years. Though more than two years have passed, someone driving through some of the devastated communities will think that we have not done any recovery work, when in actuality we have been rebuilding infrastructure, such as sewer and water, which must happen before we can rebuild the actual buildings.

You have to have the people in place who understand the potential for a disaster or a catastrophic event, and these people should know immediately how to respond and also know the management tools to use when responding, such as the National Incident Management System, which is used by all state and local governments as their method of incident management and which can be an extremely effective system if people know how to use it.

My responsibilities as director of the Mississippi Emergency Management Agency have changed dramatically over the years. While it used to be about planning for a catastrophic event while actually responding to and recovering from minor disasters, I now know what it is like to be a part of a catastrophic event, and most of my efforts are focused on the recovery of Hurricane Katrina, while still planning for the next disaster, whether it will be a tornado in the middle of the state or an ice storm in the northern part of the state.

We cannot forget, however, that we have not seen the worst. I am positive that Hurricane Katrina will not be the worst disaster our nation will face. There is the threat for catastrophic events that could exceed the devastation Katrina caused, such as an earthquake along the New Madrid fault line, which would affect several states at once.

I cannot stress enough the importance of individual preparedness. People have to realize that their emergency management officials are doing all they can to prepare for catastrophic events, but there will be a period of time when we will be overwhelmed, and it will be up to the individuals, if possible, to take care of their basic needs for the first thirty-six to seventy-two hours.

Emergency preparedness and management is not just the government's responsibility. It is the responsibility of us all.

Governor Haley Barbour appointed Mike Womack executive director of the Mississippi Emergency Management Agency in December 2006. Mr. Womack previously served as MEMA's deputy director and as the agency's response and recovery director.

During Hurricane Katrina, Womack was designated the state coordinating officer and coordinated the state's resources and the massive amount of resources provided by other states through the Emergency Management Assistance Compact. He continues to serve as the governor's authorized representative for the continuing Hurricane Katrina recovery. He has also served as the state coordinating officer for Hurricane Ivan and as deputy state coordinating officer for two additional major disaster declarations.

Mr. Womack has an extensive operations and management background and brings this knowledge to the emergency management setting. He retired in 2001 after serving twenty-nine years in active and reserve military service as a full-time lieutenant colonel for the Mississippi Army National Guard. During his time in the National Guard, Mr. Womack developed plans and exercises in support of the state's mission for emergency response. These exercises fully incorporated the National Guard's support to civilian authorities.

Mr. Womack began his career with the Mississippi Military Department in 1985 as the senior active duty officer, supervising a 1,000-man armored cavalry squadron. In 1994, he was promoted to the senior full-time officer position in the 4,000-solider 155th Armored Brigade.

Mr. Womack has received numerous awards for his military service. Among those awards is the Mississippi Magnolia Cross, which is the highest peacetime award presented by the state. He is also the recipient of multiple Army Commendation Medal awards and Meritorious Service Medal awards.

A native of Hernando, Mississippi, he has a bachelor's degree in music from the University of Mississippi and was commissioned a second lieutenant from Officer Candidate School at Fort Benning, Georgia.

A member of the Mississippi Civil Defense/Emergency Management Association, Mr. Womack is the regional vice president of the National Emergency Management Agency and serves on its Legislative Committee. He is a member of the board of directors of the Central United States Earthquake Consortium, the National Guard Associations of Mississippi and the United States, and the United States Army Armor Association, where he holds the Order of St. George, signifying significant contributions to the advancement of the Cavalry and Armor Units.

Emergency Management and Disaster Assistance in the United States and the Ramifications of the Evolution of Homeland Security

Timothy Manning

Director

State of New Mexico Department of Homeland Security and Emergency Management

The Genesis of Government Disaster Relief Programs

On a late summer day in AD 79, Mount Vesuvius erupted in southwestern Italy, first covering the region in ash fall for an entire day, and then sending a series of massive pyroclastic surges and flows throughout Pompeii, Herculaneum, Oplontis, and other towns in the surrounding areas. The destruction and loss of life were massive. In the wake of the catastrophe, the Roman Emperor Titus traveled to the area to view the damage; he declared an emergency and established a relief fund, appointing two former consuls (*Curatores restituendae Campaniae*, or "relief commissioners" for Campania) to oversee and direct the relief and rebuilding efforts.[1] Thus began perhaps the first recorded government disaster relief program. Disaster relief would be performed by governments in much the same way for the next 2,000 years, with ad hoc agencies and programs being developed in response to individual events and crafted around specific needs of the victims, survivors, and affected communities.

Disaster assistance in the United States followed much the same formula. Beginning in 1803, when an assistance package was passed for a fire-ravaged New Hampshire town, Congress would handle disaster assistance through a series of more than a hundred pieces of ad hoc legislation dealing with specific events. Agencies throughout the federal government were given emergency authorities to assist in response and rebuilding. This continued until the Disaster Relief Act of 1974 (P.L. 93-288), and the later amendments of the Robert T. Stafford Act of 1988 (P.L. 100-107), and President Jimmy Carter's Executive Order 12127, creating the Federal Emergency Management Agency (FEMA).

Government Organization

Originally created at the request of the National Governors Association to consolidate and coordinate disaster assistance and relief, FEMA also had responsibility for national continuity of government programs and, according to President Carter's 1978 "Reorganization Plan No. 3" (44 FR 19367, 3 CFR, 1979 Comp., p. 376), was assigned the mission of the federal side of civilian planning and coordination of events of "all hazards," including terrorism and war within the United States. Throughout the following twenty-five years, FEMA would continually add to its portfolio of responsibilities and programs in much the same manner as disaster relief did in the preceding 2,000 years; as new events of unprecedented magnitude or complexity would occur, new laws would be enacted, establishing new authorities or responsibilities or reforming those in existence.[2]

This constant reform continues today. In 2002, in the wake of the terrorist attacks of September 11, 2001, Congress passed the Homeland Security Act of 2002 (P.L. 107-296), which reorganized both much of the security apparatus of the federal government and its disaster and emergency response and recovery capabilities. Twenty-two different agencies were pulled together into one large cabinet department, including FEMA. All of the newly consolidated agencies' missions were refocused into terrorism prevention and response, leading to a series of internal organizational shifts, program reform, and reprioritization of resources.

Through the three years between the formation of the Department of Homeland Security and the landfall of Hurricane Katrina in 2005, many people throughout the emergency management profession warned of a potentially dangerous bias toward terrorism preparedness and response and a shift away from an "all hazards" program. Even the term "all hazard emergency management" began to be used more as a synonym for natural disaster management. An increasing bifurcation between terrorism and natural disaster programs developed, demonstrated, according to critics, by policy such as the National Planning Scenarios being promulgated by the Department of Homeland Security and the Homeland Security Council. Of the fifteen listed as the "foundational structure for the development of national preparedness standards from which homeland security capabilities can be measured because they represent threats or hazards of national significance with high consequence," only two (earthquakes and hurricanes) are natural hazards; the remainder comprise intentional attack modes.

In the wake of the perceived failures in relief efforts following Hurricane Katrina, there was another wave of reform acts. The Post-Katrina Emergency Management Reform Act (P.L. 109-295, 120 Stat. 1355, 1394) again reorganized the Department of Homeland Security and FEMA.

Relief Programs

Throughout these reorganizations and reprioritizations of mission, very little reform of the underlying disaster relief legislation occurred. While a great deal of effort was put into review and revision of plans and procedures for response to and prevention of acts of terrorism, the Stafford Act was thought to have been crafted in such a way that it would provide relief under all derivations of emergencies and disaster that may occur; however, the reality remains that the majority of the disaster relief programs used most frequently in the United States are not managed by FEMA or accomplished through the Stafford Act. These include the programs managed by the Small Business Administration and the Department of Agriculture.

For many years, emergency management professionals have suggested changes in the Stafford Act to accomplish a variety of disaster relief programs and expand authority. For example, the definition of "major disaster" includes a list of event types, but excludes public health crises, such as a pandemic influenza. Current plans for implementation of the Stafford Act include limiting declarations to an "emergency," of which there are no limitations on event type; yet this change may turn out to be insufficient to cope with the probable results of such an occurrence.

An underlying key component of disaster relief—and its frequent (or infrequent) reform—is the financial impact on the affected communities. Stafford Act relief is directed primarily at governmental organizations—a way to recoup expenditure of the local tax base by the affected communities. This is in the manner of cost match grant from the federal government to state and local governments under a program known as Public Assistance, or PA. The decision matrix for major disaster declarations is fairly objective, based on per capita damage cost threshold linked to the consumer price index. The possibility of federal disaster assistance resulting from damage incurred is moderately easy to predict, and expectation management is a fairly simple prospect.

Disaster relief to individuals and businesses comes from a Stafford program known colloquially as Individual Assistance, or IA. Unlike PA, much of this assistance is in the form of loans, rather than grants. Further, eligibility for presidential IA declarations is not determined by clear thresholds and is a wholly subjective decision at FEMA and the White House. This becomes difficult to explain or understand in cases of small-scale disasters, where damage is severe but not widespread. To homeowners or small business owners, the fact that they are not eligible for assistance because a tornado destroyed only their block as opposed to three blocks in a neighboring state is a difficult proposition to understand or accept. Frequently, in the wake of such small-scale disasters, especially ones with significant media attention, there are calls at the state level for establishment of IA programs to address these gaps.

There are few state-level individual assistance programs, and those in existence have met with varied success. Disaster relief in general began as individually enacted laws whereby the federal government would provide assistance and relief to states when they were overwhelmed. These coalesced into defined federal governmental programs and expanded to fill the field of law. States rarely created programs in this area of relief, as there was not a recognized need to augment federal law. As federal disaster assistance became more bureaucratized and regulated over the years, some states created programs to fill the gap between the status quo and the threshold where federal assistance becomes available. These actions have, however, not become commonplace. Federal policy seems to have had the effect of providing a disincentive to states' creating individual and family assistance programs. In at least one case, the existence of a state program resulted in less federal assistance because the federal government deemed that needs were already being met.

The Role of the Attorney in Disaster Response

The role of the attorney in disaster response is complex and too often overlooked. A disaster is, by definition, a condition where a "normal" way of doing things is insufficient to meet the exigency of the situation; where the rules, regulations, and laws created for checks and balances to ensure the proper expenditure of public funds and protection of civil liberties may not be sufficient to operate. Extraordinary actions may be required to save lives, including the use of measures normally considered anathema in our liberal, democratic structure, such as forced evacuations and forced quarantine. Government operations in disaster and emergency response are therefore much more complex, and potentially without precedent, and the role of the attorney is critical. There is much regulation and case law dealing with "standard" emergency response, but little in regard to extraordinary actions available to the government in times of disaster. All states, for example, have disaster statutes dating in most cases to the cold war period. These include the most drastic of the extensions of the sovereign police power of the state, such as martial law, and the suspension of constitutional rights such as habeas corpus. With the possible exception of some of the more disaster-prone states, few states have even used these powers, and few attorneys in government understand their use or limitations. The infrequency of disasters seems to lead to very little attention being paid to this sector of the law. It is unfortunate that when an attorney expert in this sector is needed most, one may not be available.

Interstate compact law is another example where the complexities of disaster response law lay outside the experience of most attorneys. Weaving the state and local law of fifty-four states and territories together with federal disaster relief law and regulations is complex at best. Interstate mutual aid provided under the Emergency Management Assistance Compact (EMAC; P.L. 104-312, statutes of the fifty-four states and territories) has become as prevalent as or more prevalent than direct federal assistance in response operations. In the response to Hurricane Katrina, more than 1,300 search and rescue assets were mobilized from sixteen states, resulting in the rescue of more than 6,000 people; more than 3,000 firefighters, and more than 6,800 law enforcement officers were sent into the area from around the country.[3] All told, more than 65,000 people and uncounted pieces of equipment were sent to the Gulf Coast from other states. The resulting web of state and federal law and disaster relief regulation has lead to complex issues that are still being adjudicated in some cases. Attorneys specializing in the field have become a rare and valuable resource to the policy maker, agency head, and governor.

Terrorism, Homeland Security, and the Evolution of Emergency Management

Counterterrorism, antiterrorism, and terrorism prevention fall under the umbrella of what, in Homeland Security parlance, is simply termed "prevention." Prevention represents a new and challenging focus for state and local governments. The United States was founded with a strong tradition of local-level preparedness for external aggression, and from the period between the First World War and 2001 (and particularly during the cold war), governmental policy initiatives and operational activities focused on external threats directed against nation states and coalitions, and were the sole purview of the U.S. federal government. States and localities worked in partnership to a large degree on continuity planning, nuclear war preparedness, and transnational drug smuggling; they did not work together, however, on transnational attack prevention. Where terrorist threats from sub-national groups and organizations did surface in domestic U.S. communities, they were, with some notable exceptions, domestic in origin, conducted by U.S. citizens, and treated as law enforcement issues. Suggestions to address the growing threat in new and innovative ways were often promptly dismissed.

The Recent Past of Terrorism Prevention and Preparedness

The 1995 sarin gas attack in Tokyo by Aum Shinrikyo resulted in the enacting of new laws and federal grant programs targeted for training U.S. domestic law enforcement and emergency responders in chemical and biological weapons attacks; however, counterterrorism operations and policy debates remained with the federal government. This state of affairs changed overnight in late 2001 with the al Qaeda attacks on New York and Washington, D.C., and, in some ways more importantly, with the anthrax attacks some weeks later. Suddenly, state and local governments were faced with the responsibility to understand a threat they had only vaguely even heard of, enact new laws and policies, and establish governmental agencies with surprising, and often unhealthy, speed.

As of 2006, most, if not all, states and localities have been through many different iterations in structure and authorities for the counterterrorism mission. There remains little debate as to who within government should be responsible for the preparedness for response to an attack and for the response itself when one should become necessary; yet the issue of who is "in charge" of preventing them remains contentious and unfocused. The U.S. federal government has undergone at least three major reorganizations (the formation of cabinet Department of Homeland Security; an almost immediate reorganization of that department—the 2SR; and the formation of the Office of the Director of National Intelligence) and numerous smaller ones, including the Post-Katrina Act reorganization of DHS, the formation of U.S. NORTHCOM in the Department of Defense, and too many more to illustrate. Through this period, the state of New Mexico attempted operation under most of the variations seen throughout the country and initiated the development of prevention programs in many different agencies of government. Since 2006, New Mexico has been aligning and organizing prevention activities as part of the National Security/Homeland Security fabric, in accordance with the National Strategy for Counterterrorism, the National Strategy for Homeland Security, Homeland Security presidential directives, example practices developed by other states, and innovative new programs developed in-house. As of July 2007, after many different variations, a cabinet agency was created to unify the efforts under a single entity with clear statutory authority.

Current Development: One State Example

Prevention, as discussed here, comprises both the efforts to prevent an attack and the steps to reduce the impact of an attack, should one occur. It is limited to detection, interdiction, deterrence, and hardening/redundancy; other areas such as training, exercising, and equipping responders fall into the preparedness area.

The overarching prevention program for New Mexico consists of a threat-vulnerability-risk analysis and a direction of resources to mitigate that risk. Current prevention activities in New Mexico revolve around two core program pillars: the collection, analysis, and distribution of intelligence; and field deterrence and interdiction operations against domestic and transnational asymmetric threats and weapons of mass destruction counter-proliferation programs. These efforts are designed against what have been assessed as the attack modalities with the greatest statistical likelihood at present:

1. Conventional explosive (IED, or improvised explosive device, and VBIED, a vehicle-borne IED)
2. Active shooter(s)
3. Suicide bombers
4. The statistically unlikely but extremely high consequence of weapons of mass destruction (chemical, biological, radiological, nuclear)

The Intelligence and Security Bureau, in cooperation with the Response Bureau and the Emergency Operations Center, is the cornerstone of the New Mexico prevention program. The I&S Bureau runs the New Mexico All Source Intel Center (NMASIC), the state's counterterrorism intelligence fusion program. NMASIC works directly with the intelligence community and conducts collection, analysis, and publication for all homeland security partners, including state and local law enforcement, emergency management, public health, and the private sector. While they focus on terrorism and national security intelligence, as it is often impossible to know where one stops and another starts, they work in cooperation with all agencies in all-crime analysis, to include organized crime, drug cartels, outlaw motorcycle gangs, and street gangs. They have recently completed the first-ever all-source threat analysis, an NIE (National Intelligence Estimate)-scale review of the groups posing a threat to New Mexico, their intent and capability to conduct an attack or harm our communities, and an overall assessment of the risk posed by those groups. This report will be used in conjunction with vulnerability analyses to drive decision-making in spending, resource allocation, and program expansion/revision.

New Mexico was one of the last states to develop a permanent intelligence center or capacity. Homeland security is really a matter of information sharing and involves bringing states and local governments into the intelligence community so information can be shared and puzzles completed. Its goal is to establish the capacity to share information among local governments, hospitals (symptomatic information), state agencies, and the federal government. As has been later discovered, much of the activity of the 9/11 conspirators was known to individual agencies and levels of government, but there was no sharing mechanism in place. Intelligence operations at the state level now allow us to do this, providing useful information to law enforcement, executives, and governors on a regular basis. It also allows for strategic topical research, such as trends in emergence threat groups, to allow a more informed use of limited funds and budgeting decisions. Daily product would be produced for decision-makers—or more frequently, as necessary. Our national intelligence system was developed throughout the cold war to address a specific nation-state threat, and it is unsuited to the asymmetric, sub-national/transnational threat that we currently face. State- and local-level national security intelligence operations are a natural outgrowth, an organic development in response to the threat as perceived by state and local governments, and their feeling of responsibility toward action.

I&S also runs the Critical Infrastructure Protection program, where, in accordance with New Mexico's Homeland Security Act, analysis of vulnerabilities and work toward target-hardening of critical facilities, sectors, and locations and standardizing facility security standards for governmental buildings are conducted. In cooperation with the Radiological and Hazardous Materials unit, they identify the most dangerous chemical storage and stockpiles in New Mexico, and work toward hardening these locations.

Border Security Programs at the State Level

Another high-profile prevention program in many states is security of the international border. Article I of the Constitution and the Tenth Amendment outline those specific responsibilities of the federal government (enumerated powers) and those reserved for the states (reserved powers). Border security and immigration has always been the sole domain of the federal government, and law enforcement in our communities the province of the states and their political subdivisions. In the wake of the 2001 attacks, as well as the constant fear of follow-up attacks, the federal government was widely perceived as failing in its responsibility on border security. Criticism of federal immigration and border security policy has been common and cyclical for many years, usually tied to the general economic health of the United States, and most specifically unemployment. Unlike the usual criticism, this sudden outcry against federal policy occurred at a time of very low unemployment and was based primarily on a feeling of vulnerability. Governors throughout the Southwest committed funds to the problem, bolstering local law enforcement; meanwhile, the federal government was perceived to be reducing or eliminating its support.

The security situation posed in the border regions resulting from an immigration policy that does not take economic drivers, such as low-paid construction jobs, into account, as well as from border control tactics that involve the virtual abdication of ground to afford a layered-zone defensive system, required non-traditional immigration and border agencies to become involved. Through the granting of state, as well as federal, funds to local law enforcement, criminal interdiction and crime deterrence operations are conducted throughout the border region by state and local law enforcement.

Emerging Prevention Issues

To date, prevention programs in response to terrorism have been largely extensions, expansions, and modifications of existing programs. In the rush to build a homeland defense and security system, the development of prevention activities, for the most part, mirrored the past experiences and contextual framework of the individuals in policy roles at the time of formation. Specifically, homeland security policy-makers with a career in the military designed and enacted programs decidedly military in nature, for which they used doctrinal frameworks borrowed from the military but largely unknown and misunderstood by the civilian world; a policy-maker pulled from a life in civilian law enforcement tended to look at it as a law enforcement problem, and programmatic developments mirrored traditional criminal justice activities; public health officials focused on bioterrorism to the exclusion of other threats; emergency management officials tended to look at it as just one more hazard, and so on.

Government functions in existence in early 2001 were created and evolved through time in response to the needs of the public as perceived at that time. A plausible threat of sufficient magnitude to potentially disrupt our system of government, or even bring down the republic, was not something widely acknowledged, and no existing government structure or trans-governmental policy construct was in place to deal with it.

Dispositional Prevention

Most prevention programs in the states are centered on operational deterrence and interdiction, which include activities to detect and prevent actions by those already radicalized and determined to take action. Many recent terrorist attacks in Europe and plots disrupted in North America have involved natural-born citizens of Western countries. A comprehensive radicalization prevention program is often discussed and greatly needed. To date, there has been no coordinated, widespread effort, beyond occasional discussion, to enact programs to attempt to prevent the radicalization of individuals, or, if radicalized, to prevent them from taking the last step of deciding to act on their beliefs.

Another often-discussed dispositional prevention program option is a practical increase in international disaster assistance, such as additional financial aid and direct assistance from the Department of Defense, the Office of Foreign Disaster Assistance, or even collections of state and local assets sponsored by the U.S. government to assist in crisis management and disaster response. U.S. Pacific Command's response to the Indonesian tsunami resulted in U.S. government approval ratings by people of the region never before seen, although these numbers have since declined in the face of constant news of other sorts from the region. While the factors leading toward radicalization are numerous, the appearance of the United States and the West in the eyes of the world is a consistent feature. Domestic and international prevention strategies, however, cannot be limited to the interdiction and apprehension of existing actors, but must include the prevention of their genesis.

The Near Future: What Is Next?

Based on historic trends, it is clear that no more than a few years go by without some reform or reorganization in emergency management. It is likely that this trend will continue. Furthermore, as the profession has been evolving over the past seven years, it is likely to evolve into something decidedly different from either what is currently seen as homeland security or "traditional" emergency management. It is likely that a new profession is developing, or at least a new schema for organization of the same governmental tasks that have been performed invariably over the past 2000 years.

As civil defense has grown from civil preparedness and back into emergency management, we are at a cusp of organizational development into a new field, shedding the paradigms and dogma of the recent past and moving into a method for organizing the reduction of natural disasters and the effects of global health crises and power shortages, as well as addressing the prevention of terrorist attacks and the response to and recovery from the effects of all of these.

1. Gaius Suetonius, *The Lives of the Caesars, II Claudius. Nero. Galba, Otho, and Vitellius. Vespasian. Titus, Domitian. Lives of Illustrious Men: Grammarians and Rhetoricians. Poets (Terence. Virgil. Horace. Tibullus. Persius. Lucan). Lives of Pliny the Elder and Passienus Crispus.* Edited and translated by J.C. Rolfe (Cambridge: Harvard University Press, 1914).
2. Clair B. Rubin, Irmak Renda-Tanali, and William Cumming, *Disaster Timeline: Major Focusing Events and U.S. Outcomes (1969–2004)* (Arlington, Virginia: Clair B. Rubin and Associates, 2004).
3. Emergency Management Assistance Compact, *2005 Hurricane Season Response After-Action Report* (Lexington, Kentucky: National Emergency Management Association, 2006).

Timothy Manning is the director of the New Mexico Department of Homeland Security and Emergency Management and homeland security adviser to Governor Bill Richardson. Mr. Manning was named the department's first director by Governor Richardson in April 2007, having previously been appointed to the cabinet as director of the Governor's Office of Homeland Security in 2005 and as the state director of Emergency Management since early 2003. He has also served in the Richardson Administration as a deputy cabinet secretary of the New Mexico Department of Public Safety.

With a diverse background in emergency services, Mr. Manning has worked in a number of positions in the state's emergency management agency, including chief of the Emergency Operations Bureau, and has been, at various times, a firefighter, an emergency medical technician (EMT), a rescue mountaineer, a hazardous materials specialist, and a hydrogeologist.

In addition to homeland security, Mr. Manning oversees the daily administration of the state's disaster and emergency preparedness, mitigation, response, and recovery efforts. In addition to his role in the cabinet of Governor Richardson, Tim is a guest lecturer and subject matter expert at the Center for Homeland Defense and Security at the Naval Postgraduate School in Monterey, California.

Mr. Manning is chair of the state's Emergency Response Commission, Intrastate Mutual Aid Commission, and Homeland Security Advisory Committee. At the national level, he is currently co-chair of the National Homeland Security Consortium, chairman of the

National Emergency Management Association's Homeland Security Committee, and chair of the Emergency Management Accreditation Program (EMAP) Commission, an international emergency management standards setting and accreditation body. He has previously served as the chairman of the Response and Recovery Committee and regional vice president of NEMA, and continues to serve on many other national and state boards and commissions.

Mr. Manning received a Bachelor of Science degree in geology from Eastern Illinois University, is a graduate of the Executive Program at the Center for Homeland Defense and Security of the Naval Postgraduate School in Monterey, California, and is currently doing postgraduate work at the Center for the Study of Terrorism and Political Violence at the University of St. Andrews, Scotland.

Recovering from Disasters: Learning from Domestic and International Experiences

Carlos J. Castillo
Emergency Management

I began my public service career with the Miami-Dade County (Florida) Fire Rescue Department in 1981. As a firefighter and paramedic for one of the nation's largest and busiest fire and EMS departments, I experienced firsthand the human condition and the ways people dealt with life-threatening emergency situations.

When I became involved with disaster management internationally in the 1980s, I began to realize the similarities that exist in how affected populations deal with disasters. Many developing nations seemed to have discovered the benefits of mitigating the effects of and preventing disasters. However, communities, elected officials and even some emergency management professionals place more emphasis on preparing to respond to disasters than preventing them until they have actually been affected directly by them.

The recovery phase of disaster management is arguably the least understood of the disaster cycle, which includes preparedness, response, recovery, and mitigation. Emergency management professionals at all levels of government spend the majority of their time and efforts on preparedness and response activities. Catastrophic disasters occur infrequently compared to the majority of emergencies. Therefore, most areas lack the experience to proactively address long-term recovery challenges.

Disaster Housing

In 1988, a devastating earthquake struck Soviet Armenia, and Miami-Dade Fire Rescue was called on by the U.S. government to provide search and rescue. I responded as part of the team, along with members of the Fairfax County (Virginia) Fire Department. We recognized that much could be learned from the international response that drew humanitarian assistance from throughout the world. Following the response, the United States realized the need to develop and improve its capability to manage a major disaster in a large metropolitan area.

The Armenian earthquake resulted in more than 25,000 deaths and many thousands of homeless. Tent cities and other makeshift shelters sprang up throughout Leninakan and other cities.

On August 24, 1992, Hurricane Andrew, only the third Category 5 storm to make landfall in the continental United States, decimated the southern half of Miami-Dade County. The county was faced with having to find shelter for thousands of people who were left homeless overnight. Tent cities sprang up throughout Homestead and Florida City.

In September 2005, Hurricane Katrina struck the U.S. Gulf Coast, killing more than 1,500 people and leaving scores homeless. At its peak, FEMA (the Federal Emergency Management Agency) housed almost 150,000 families in manufactured housing and travel trailers. Two-and-a-half years later, thousands remain in mobile homes and trailers. Available housing in the affected states is virtually nonexistent, and FEMA, the Department of Housing and Urban Development, and state and local agencies are working to find more appropriate housing for the 30,000 families remaining.

From 2003 through 2005, I was the director of Emergency Management for Miami-Dade County. In 2005, Hurricane Wilma struck South Florida. Many homes had already been rendered uninhabitable by Hurricane Katrina two months earlier. Wilma added insult to injury. Once again, county officials were faced with trying to find shelter for our displaced population. We worked to find adequate housing for families, as opposed to creating tent cities. Many challenges arose, since affordable housing in Miami-Dade was scarce, and hundreds were on waiting lists for subsidized housing.

For more than ten years, I managed a Miami-Dade fire rescue agreement with the U.S. Agency for International Development. The grant-funded program focused on responding to disasters on behalf of the United States throughout the world, and on preparing first responders in Latin America and the Caribbean to be self-sufficient in disaster response. I led teams from Miami-Dade to most major disasters around the world that required search and rescue where the countries requested humanitarian assistance from the United States. We worked with our counterparts in the region to develop disaster preparedness training programs for first responders.

Working with colleagues from around the world, it became evident to me that although some of our disaster threats and vulnerabilities differed from theirs, we had much in common with them and their approach to disaster preparedness and response. Similar to domestic disasters, economically disadvantaged populations invariably take the longest time to recover from catastrophe.

I also quickly realized that there were almost as many cultural and socioeconomic differences among Latin American countries as there were between them and the United States. To be effective, preparedness and response activities must be adapted and tailored to the target population.

Preventing Disasters

I do not believe there should be a distinction between natural and man-made or, more appropriately, human-caused disasters. Earthquakes, hurricanes, and other weather-related events need not result in disasters. Although we cannot stop the wind from blowing or the earth from trembling, we can prevent these natural phenomena from causing harm to people and property. Human inaction contributes as much to disasters as human action. Designing structures that can withstand the threats they may face, coupled with effective building code enforcement, will keep people safe and buildings secure.

The majority of the U.S. population lives within one hour of a coast. Building in areas that are prone to hurricanes requires stronger and thus more expensive construction and materials to ensure safety.

Arguably, regulating land use with the goal of preventing development in areas prone to earthquake, flooding, and other natural hazards is a major step toward preventing disasters. However, these regulations are usually opposed by developers, private landowners, and local elected officials.

Homeland Security and All-hazards Planning

In the post-9/11 era, the term homeland security has evolved to encompass more than terrorist threats. Federal, state, and local governments placed much attention to preventing human-caused disasters and the need to increase awareness of potential threats of terrorism. However, the extremely active hurricane seasons of 2004 and 2005, especially Hurricane Katrina in the Gulf, provided a wake-up call for many and have served to refocus attention on preparedness, prevention, and mitigation of all types of disasters. This approach is referred to as all-hazards planning and preparedness.

All disasters are local. When disasters occur, the local community is the first on the scene, and is in the best position to save the greatest number of lives. As the response to the disaster evolves, the focus shifts toward disaster assistance and recovery. Most disasters and emergencies attract overwhelming media attention in the hours and days immediately following the event, which in turn results in an unanticipated outpouring of assistance from individuals, agencies, and volunteer groups. As time passes and the outside attention subsides, the local community must face the long-term challenges of rebuilding their homes and their lives with undoubtedly less help from outside their community.

One of the top priorities for the U.S. Department of Homeland Security and FEMA is to foster a "culture of preparedness." FEMA is working to create and foster partnerships at local, state, and federal levels. Citizen Corps Councils and Community Emergency Response Teams provide the opportunities for citizens to prepare for natural and human-caused emergencies.

At the city, county, and state levels, mitigation, preparedness, response, and recovery activities are planned, based on community threat assessments and vulnerability analyses. The result is a focused approach to the emergencies they are most likely to face and pre-designated roles and responsibilities.

In the past, state and federal agencies waited until the local jurisdiction was overwhelmed before beginning to provide assistance. Recently, the approach is to "lean forward" and work closely with local jurisdictions as they prepare for and respond to disasters, not waiting for sequential points of failure before responding.

Faith-based organizations, such Catholic Charities, Church World Services, and the Southern Baptists, play a major role in providing humanitarian assistance to individuals affected by disasters. They also provide a natural venue for dissemination of disaster-related information.

Many states have realized the important role played by volunteer individuals in providing disaster relief. Volunteer Florida, created by the Florida Legislature in 1994, encourages volunteerism for disaster preparedness and response. It is the lead agency for the coordination of volunteers and donations at the Florida Emergency Operations Center following a disaster.

Elected officials at all levels of government, especially at the local level, are expected to assume a lead role in directing response to and recovery from disasters. Following the 9/11 terrorist attacks on the World Trade Center, New York City Mayor Rudy Giuliani exhibited strong leadership that clearly contributed to how the city and the rest of the country coped with the tragedy.

Much has been written on the response to Hurricane Katrina in the Gulf. Undoubtedly, the challenges faced at every level of government were unprecedented. The U.S. Congress passed the Post-Katrina Emergency Management Reform Act in 2006 to ensure that government in fact learns from the response and implements necessary changes to prevent a recurrence.

Increasingly, the public is holding elected officials accountable for their actions in response to disasters in their community. Regrettably, in most cities, this happens only after an event that results in major property damage or lost lives and unnecessary suffering. Accountability must start with *preventing* disasters. Too often, the public or even emergency management professionals realize the benefits of prevention after they've been affected.

Who should pay when you are affected by a natural disaster—you, your insurance company? Should your city, county, or state be responsible for reimbursing your uninsured expenses to restore what you had before the event? Should the federal government be responsible for making you whole if you are not adequately insured and/or the state you live in has no programs in place to assist victims of disaster?

Approximately 90 percent of all disasters in the United States are flood-related. In 1968, Congress established the National Flood Insurance Program (NFIP). NFIP was created to provide affordable flood insurance to property owners when coverage through the private insurance industry was too expensive or difficult to obtain. Only property owners in communities with floodplain management programs are eligible, thereby theoretically limiting vulnerability.

Attorneys can play a vital role in natural disaster relief. Protecting citizens' rights when they are most vulnerable following a disaster is essential to enabling to people to return to normalcy. Trying to understand the complex insurance policies that are constantly changing and becoming more restrictive in what is covered is overwhelming.

Local and state officials need guidance in navigating the laws and regulations that govern disaster relief. In major disasters, where federal assistance may be required, attorneys play a key role in advising state and local officials whether to request a presidential disaster declaration and what to ask for.

Changes and Improvements

Most major disasters result in a multitude of lessons learned, and thus rules and regulations are revised and updated to reflect these lessons. An example is the Post-Katrina Emergency Management Reform Act of 2006 (PKEMRA) (P.L. 109-295), which made substantial revisions to the Robert T. Stafford Disaster Relief and Emergency Assistance Act (42 U.S.C. 5121 et seq.) (PL 100-707), the law governing disaster response. PKEMRA also reformed national preparedness and response to disasters; improved disaster planning, including the evacuation of people with disabilities and other vulnerable populations; improved preparedness and training; and implemented new measures to prevent fraud, waste, and abuse during emergencies.

Current issues in natural disaster relief affect most government systems. The National Response Framework, a guide to how the United States responds to disasters, describes the roles and responsibilities of entities involved in managing emergencies at the local, tribal, state, and federal levels, and their interaction with the private sector and non-governmental organizations.

Naturally, the U.S. Department of Homeland Security and FEMA have been affected by lessons learned in past disasters, especially Hurricane Katrina and the 9/11 terrorist attacks. However, the changes have affected virtually every level of government. The private sectors, as well as government agencies, have realized the importance of planning for the safety, health, and welfare of their most important asset—their employees.

Timely and necessary pressure on elected officials and emergency managers to protect their constituents has been a powerful force in bringing about necessary changes in how we prepare for, respond to, and recover from disasters. Changes to the way the federal government manages disasters are most often congressionally mandated. Local and state governments enact changes based on their disaster vulnerabilities or as directed by federal laws. Unfunded mandates, however, pose challenges for state and local governments, especially those that may not have been directly affected by disaster, and emergency managers must create awareness of the importance of disaster preparedness. Federal preparedness grants to states and local governments help alleviate the financial burden. Hazard mitigation and pre-disaster mitigation grants provide funding for smarter rebuilding to prevent future disasters.

FEMA created the National Advisory Council and Regional Advisory Councils to solicit input from other local, state, and federal partners, as well as the private sector and non-governmental organizations, in policy development. It is felt that input from those who will most likely be affected by policy changes leads to a better product that is more likely to be accepted at all levels.

The first step to acceptance of changes to disaster preparedness and relief programs is creating awareness of the need to be prepared. Unfortunately, most people heed preparedness advice only after they experience the effects of a major disaster on their community or neighboring jurisdictions.

The effect on citizens of changes to disaster relief programs is that they are more aware of threats to their communities. Fostering a culture of preparedness means we are all in this together. Preparedness and awareness must begin at the local level. The result is saving lives and the reduction of physical and emotional pain and suffering.

Federal funding for fire departments was virtually non-existent prior to 2000, when Congress created the Assistance to Firefighters Grant Program to address the fact that too many fire departments in this country lack even the most basic needs, including proper turnout gear, training, communication systems, and prevention and public education programs. By providing funding to local fire departments for these items, the program helps the nation's fire and emergency services achieve a baseline level of readiness.

Following the terrorist attacks of September 11, 2001, Congress established the Urban Areas Security Initiative to provide funding to state and local emergency services departments, including law enforcement, fire, and emergency medical services. The goal is to help agencies be better prepared for the threat of terrorist attacks at the local level.

Since Hurricane Katrina, the federal government has adopted a "forward leaning" approach to disaster response. Whereas in the past, state and federal agencies waited until the local jurisdiction was overwhelmed by the disaster before stepping in to provide assistance, the current philosophy is not to wait for sequential points of failure. Rather, by partnering with state and local governments and establishing relationships before an event, local officials are less likely to hesitate to ask for outside assistance to help manage the disaster. This progressive approach proved essential to managing the Southern California wildfires in December 2007. The response was a textbook example of agencies at all levels working together toward a common goal.

Training programs and improved communication, especially through use of the Internet, facilitate the implementation of changes and improvements among federal, state, local, and nongovernmental partners in disaster relief. The National Response Framework (www.fema.gov) provides necessary guidance to all levels of government and their partners. It is written as a document that can be used by elected officials in addition to disaster management practitioners.

The Internet is perhaps the most effective means for people to navigate the changes affecting disaster relief. In addition to providing information to emergency managers, disaster victims can also find essential information and guidance on what assistance programs are available to them.

Managing the expectations of disaster victims is arguably one of the key challenges to effective disaster assistance. There appears to be a misconception that disaster relief programs are intended to make people whole. In reality, disaster assistance at the federal level is intended to supplement assistance from local, state, and nongovernmental agencies. In 2008, federal assistance to individuals is limited to $28,800 per disaster. The amount is adjusted annually. This amount rarely serves to make someone whole.

It is unrealistic to expect that one agency or even one level of government can provide for all the needs of everyone following a disaster. We must therefore keep in mind that preventing disasters is inherently more cost-effective than responding to them.

Carlos J. Castillo became assistant administrator for the Disaster Assistance Directorate in July 2007. He came to FEMA after more than twenty-five years as a firefighter and local emergency manager.

In his previous position, as the assistant fire chief for Technical Services in Miami-Dade, Florida, he managed the divisions of Fire Prevention, Emergency Medical Services, and Training and Safety. In this role, he also served as the incident commander during the 2007 Super Bowl. Mr. Castillo also worked as the director of the Miami-Dade County Office of Emergency Management, where he managed the response during seven hurricane activations and oversaw Domestic Preparedness and Community Outreach and the county's participation in the Urban Areas Security Initiative program.

Mr. Castillo also has served in other assistant fire chief positions, including assistant fire chief for operations, where he worked closely with FEMA as the incident support team leader for FEMA's Urban Search and Rescue (US&R) response at the Pentagon following the September 11 attack. He also worked with FEMA earlier in his career as part of the Incident Support Team that responded to the Oklahoma City bombing. He is a FEMA US&R Incident Support Team Working Group member and a working group member for development and implementation of FEMA's US&R Response System. He was appointed to Florida's Domestic Security Advisory Panel, created following the September 11 attacks to provide advisory assistance to the governor, the legislature, and other entities.

Mr. Castillo earned a bachelor's degree from Barry University and a master's degree in public administration from Florida International University. He has studied at Harvard University's John F. Kennedy School of Government in the "Crisis Management: Exercising Leadership in Extraordinary Times" program.

Dedication: *I wish to dedicate this chapter to my wife, Mary, and our children, Jessica and Anthony, for their unending patience and understanding throughout my career, especially when I've had to leave home at a moment's notice to respond to an emergency anywhere in the world. To my parents, thank you for instilling in me the importance of always doing the right thing through integrity and a strong work ethic. Finally, I extend my sincere gratitude to my dear friends and colleagues in disaster and emergency services, particularly Paul Bell, my mentor.*

The Nuances of Natural Disaster Relief Issues

Dan W. McGowan

Administrator/Homeland Security Adviser

Montana Disaster and Emergency Services

Issues

Natural disaster relief programs are extremely well developed. The expectation of when anyone should receive that relief is relative to the mechanisms in place for dealing with the situation. The culture for dealing with these situations varies geographically.

Here in Montana and the West, people are pretty self-sufficient. They work with their neighbors and do what they need to do to get through the bad day; whereas, in other areas, that culture is not as prominent, and people expect the outside assistance: "Who is here to help me get through this, because it is not fair?"

If you were to compare some of the situations that you see, like during Katrina, to the huge snowstorm last year that devastated Colorado, you would see a totally different response. Am I saying that we should not have had a response to Katrina? Absolutely not. There were many unmet needs; the infrastructure was severely damaged; and the population was devastated. A presidential declaration was the appropriate response. On the other hand, the Colorado snowstorm was just as devastating to those in the middle of it.

The difference was the culture and the coordination of the mechanisms available to remedy the situation as best as possible. The culture of a given area, the available mechanisms, and the expectations drive the type of response and recovery that occurs.

The word "presidential" has its connotations. In some recent wildfires, the word "presidential" came up. The immediate expectation of individuals was, "Oh, good, we will get help. There will be a presidential declaration, and we will get help." The context of its use was totally different: damage information was being collected to see whether any type of "presidential" assistance was possible. That devastated some people because they thought the presidential disaster declaration meant assistance, and they would get the help to solve their financial problems caused by being out of work because of the disaster.

We used a town hall meeting to level the playing field and help manage the expectations. Citizens were educated about the process, the operational parameters, the types or levels of assistance we could provide, and what they could expect. Leveling the playing field brought a sense of reality to the expectations. Even though the information shattered some of their hopes, the open and honest communication was appreciated and gave the individuals a new sense of direction that was accurate and well founded. It took a lot of coordination and input from those affected, but we helped get those people the assistance that we could, and to them that made the difference.

People tie different things to different words, and when they hear "presidential declaration," I believe they have an expectation of what that means, and that will help them get through everything and they will all be taken care of. What they do not understand is that there is not a recovery program that will bring anybody back to what I would call "normal," or where they were before it happened. It will just provide the necessary assistance to get them back on their feet with regard to serious and necessary disaster-caused needs. The rest of the work to fully recover is up to them.

One thing I think needs to be clarified: it is not just about natural disaster relief, but about relief for disasters in the context of all hazards. It does not make any difference whether the disaster is from terrorism, an earthquake, or something that is man-caused like a fire—the relief mechanism or the consequence management portion of it still needs to function to provide the response or recovery services that are necessary to handle any of those events. Policy makers need to consider programs from a more global perspective instead of in a targeted or myopic fashion. With a lot of the funding that came down from Congress after 9/11, they basically said, "We believe in the all-hazard approach. You can leverage these funds for any hazard as long as there is a terrorism nexus." The reality is that terrorism is just another one of the hazards that we may face in the context of an all-hazard setting. The funding should not be targeted; it should be more what we need to do to be prepared in the event of any kind of disaster following the all-hazard philosophy adopted in the National Preparedness Goal and Homeland Security Presidential Directive 8.

I think the biggest issue in disaster relief is preparedness. The local, city, state, tribal, federal, voluntary, and non-profit organizations, as well as the citizenry, need to be prepared to the nth degree to lessen the impact of any incident, emergency, or disaster. Preparedness to me is huge, because the more prepared you are to respond to, recover from, or mitigate the impacts of any kind of a disaster, the better the chances you have of lessening the impact of that disaster when it hits. The value-added benefit of investing in preparedness is that the financial need for recovery is drastically reduced, and the devastation to those affected is minimized. Disaster history indicates that for every dollar invested in mitigation, four dollars are saved in response and recovery efforts.

The second thing that has a huge impact on disaster relief is managing effective coordination and expectations among the providers. It is so important that all the entities partnering to respond and recover have a keen understanding of each other's roles and responsibilities. Clarity in this area improves the effectiveness of the engagement by the players, reduces duplicative or counterproductive efforts, and helps minimize unnecessary financial expenditures. They must have the integration done for the inter-coordination, the communication, and the collaboration so that they can provide the best service possible in a collaborative fashion that is mission-based. It is a team approach and needs to be a team effort.

The other piece of that, at all those levels, is understanding the expectations—knowing what kinds of assistance are out there and how the system works and integrates with other delivery mechanisms. Recipient expectations tend to throw a curve ball to the providers by venturing off into the political side of things—when citizens call Congress and express their dissatisfaction with the available or non-available assistance. Whether the expectation is real or perceived, the call can sometimes impede the service delivery in order to answer the Congressional inquiry. So it is important for all stakeholders, including Congress, to understand the pieces and the parts and how they work.

Understanding the parameters and being on the same page with the partnering stakeholders improve the consistency of the information being relayed to the citizens. The coordination and collaboration aspect alleviates the potential for the citizens to question the service delivery as they would when they hear a different response every time they turn around. The important aspects are understanding the delivery mechanisms, providing consistent information, and directing those affected to the right place the first time. That approach enhances the credibility of the providers enormously in the eyes of the recipients.

The third dilemma is managing the expectations of the citizens. The emotional health of the citizens can be affected by unrealistic expectations. The citizens need to know what to expect. I have worked presidential declarations and have been in emergency management for seventeen years. I have dealt with individual assistance issues a lot. The expectations of the citizens can become so overwhelming, because of a lack of understanding, so quickly that you end up doing more work than necessary and implementing unnecessary damage control measures. If the citizens do not understand what to expect, or there is not a system in place to get the information right the first time, they end up knocking on several doors, dismayed by the closed doors because they cannot find what they are looking for. That puts them on an emotional roller-coaster ride, which you cannot afford to do. During those times you may be out of your home; everything is disjointed; and you are dealing with something that is totally out of the norm, so you are trying to normalize something that is very abnormal. You just want to get back to life as it was. So the more we can help them get directed to the right assistance providers, understand what they can expect, and help them identify justified needs, the less you have of that emotional roller-coaster ride.

During the wildfires of 2000, we put together public education pieces, and we conducted several town hall meetings. We explained to the citizens right up front, "We cannot promise you anything, but you can be assured that as far as our responsibility is concerned, we will knock on every door that we can to open up all the avenues of assistance that are possible, to help you get through it. Our goal is to make sure that nothing falls through the cracks."

We also helped them understand they have some responsibilities themselves. Insurance is the first line of defense after a disaster. Citizens need to contact their carriers to discover the coverage that may be applicable. Our part is helping them understand that there are resources to help them sort out the insurance discrepancies if they feel their policy is not being applied appropriately. So I think that managing expectations is a huge portion of one of the issues that impact disaster relief.

Fourth, you have to be sensitive to cultural distinctions and issues. You have to understand the distinctions to effectively craft the service delivery mechanisms. A prime example relates to a disaster involving a Native nation. They have their distinct and different culture, just as we have our culture. You have to be cognizant of those cultural distinctions and differences so that you can help them get through their bad day and, once again, manage those expectations. You just have to be sensitive to their needs.

For example, the Native nations expect a nation-to-nation relationship: Native nation to the federal government. The federal government through the Stafford Act, however, requires the Native nation to access any available assistance through the governor of the state where the reservation located. It is important to work with a federal Native nation liaison when working with the affected reservation. The liaison is a Native who understands the cultural differences. The Native nation also identifies with the liaison, and a sense of credibility exists from the beginning.

So preparedness, managing expectations, and sensitivity to the internal/external and cultural issues are the three major elements that always belong in the picture with the collaborative efforts providing relief during disasters.

I think some of the things that are changing in disaster relief are some of the mechanisms. Program or assistance providers realize that there need to be stronger ties between the stakeholders and the entities to enhance disaster relief. Major events like Katrina and 9/11 are agents of change to make you take another hard look at the system, the process, or the mechanism to see just what is relevant, necessary, and prudent considering the circumstances. Gaps still exist that need to be rectified to improve what we can or cannot do toward enhancing collaborative partnerships and cooperative relationships.

Look at the 9/11 commission report. The report identified that the intelligence-sharing environment was horizontally broken and vertically fractured. Agencies were operating within the boundaries of their jurisdictions lacking internal cooperation and seemingly not sharing information collaboratively across disciplines. The key to success will be to identify the gaps establishing an evaluation and corrective action development plan. A lot of progress has been made in this arena with the Intelligence Center concept. The challenges that lie ahead pertain to the integration among local, state, and federal agencies. Some of the foundation elements among and between the various provider levels are emerging as the development continues and will eventually enhance the intelligence capability.

I think other things are always in consideration and centered around funding. Do we have the right funding levels to get programs put in place? Those are always prominent issues because there is always the issue of sustainability, as well. There is only so much money to go around, so the priority of how you do things also affects those issues.

From the after-action reports with Hurricane Katrina, we are seeing a new, revitalized Federal Emergency Management Agency, FEMA, coming on board. Different changes have occurred to try to improve the delivery system. The most notable revision is the new National Response Framework and a clearer definition of the Emergency Support Functions. FEMA has worked hard to improve the deployment of response resources. The latest positive change is the ability of the FEMA director to report directly to the president in the event of a disaster; this streamlines the necessary communication channels to prevent impeding timely assistance delivery and decision making. It is a continuous improvement cycle between preparedness response and recovery and mitigation. There are many cause-and-effect relationships and issues that occur that have an impact on the system and the ability to provide natural disaster relief. It needs to be looked at from a global perspective, so all of those factors can be taken into account to try to grasp the best way we can mold the system to provide the necessary and best assistance in preparing our nation to deal with all hazards.

Components

A diagram that we use is a pictorial of emergency management, depicting the local, state, and federal levels. From a state perspective, we are in the middle because we work with assisting local and tribal nations for disasters and with their partners, not only on a day-to-day basis, but also during the disaster. We work with all of our state and federal partners to implement potential assistance mechanisms and programs. On the federal side, that is not just FEMA, but also many of the federal agencies, so that spider web keeps branching and branching and becomes very convoluted.

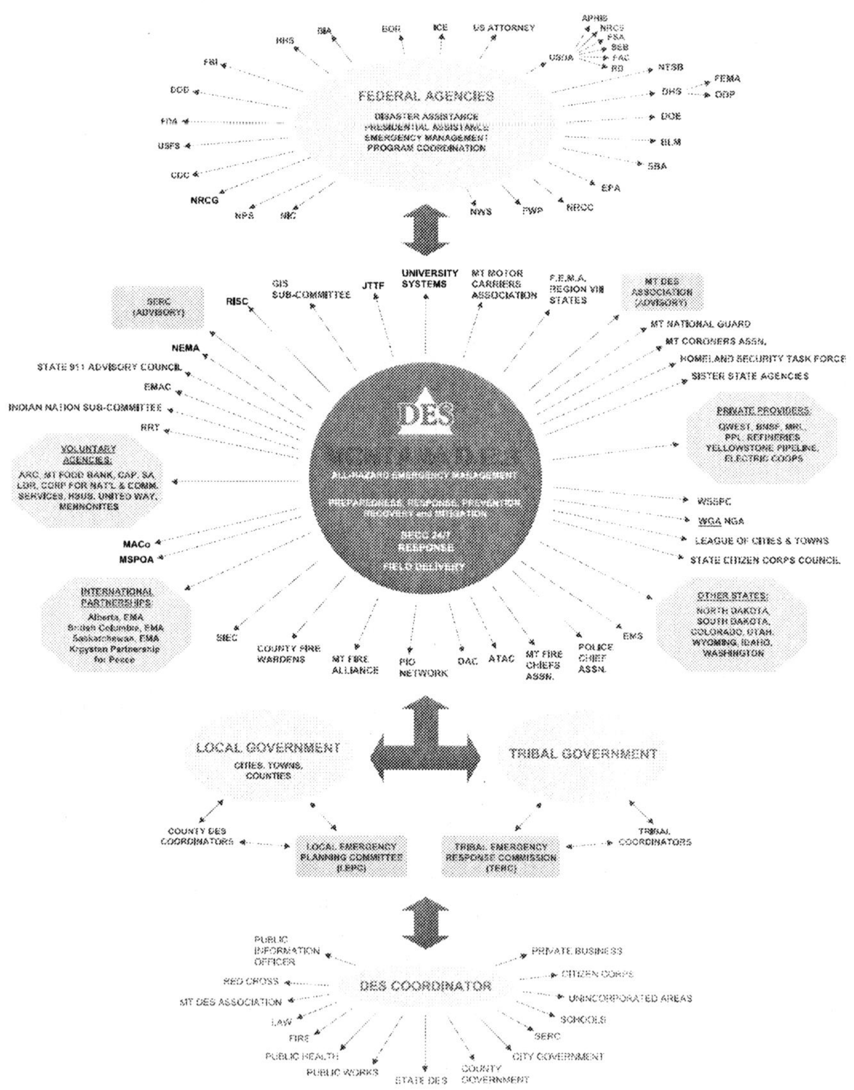

That is what makes emergency management challenging and such an exciting profession; we reach out and touch so many partners and stakeholders who are members of the team in helping to put together the assistance necessary for disaster relief. As a profession, we reach out and touch more agencies and more partners than probably any other profession in the country to do what needs to be done to provide disaster relief. If we are not the most, we are one of the most diverse professions based on the multitude of entities we cooperate, coordinate, and collaborate with to provide service delivery.

The emergency management mechanisms are part of a complex adaptive system. We start with pure chaos after a disaster morphing toward a controlled environment through a sense of emergence. For measuring the benefit of all that we do, it is hard to develop a metric to satisfy financial scrutiny because no disaster is ever the same. The parameters and elements are always distinctly different, and the cause-effect relationships are never the same. Besides, disasters know no geographic boundaries, and Mother Nature certainly doesn't weigh the odds to determine the best outcome. Every disaster is different. That is why it is tough to put together any kind of a control from a scientific standpoint or identify a measurable improvement between disasters.

What we do is like a life insurance policy. A policy is only as good as what you invest in, and if you do not invest very much, when you need it, then it is not going to provide you with as much as it can, so one of the keys that affect disaster relief is the investment in building a strong emergency management program from all levels: federal, state, and local.

The relief you will be able to provide is only as good as the investment and the partnerships you develop to effectively implement the assistance that can be there. In some cases where you have coordination that takes place, and it is only a one-quarter-time person doing it, realistically you cannot even open the mail, let alone do preparedness. If you send someone to training for a week, then that person is done for the month, so funding is another real issue. That is huge.

The mindset change is another major issue toward disaster relief, because that is tied to the funding piece, and the idea that, "Oh, it will never happen here," or, "We will never have to deal with that." I think if you go back and take a look at the Katrinas and 9/11s and those kinds of things, it can happen here. We cannot say when, but there is a guarantee it will happen, and it will be of a different severity or magnitude each time it happens. The main point is we all have to gather together and partner together to do what we need to do to help our citizens get through the bad day. The citizens are an important player in this mega-team effort to me, and I do not think we have scratched the surface yet.

Financially speaking, there are only so many recovery programs, like some of the disaster loans and the Stafford Act. The financial element is always under scrutiny, because you look at the money that was allocated to the response and recovery for various disasters over the years, and then you consider, "Do we put the money in on the front end of this thing, or do we put the money in after it happens?"

My point to you is that it is prudent as financial managers that we put the money in on the front end because the systems can be well oiled, the preparedness can be as cohesive as possible, and that will lessen the impact of what needs to be put out on the other end in the response phase. We need to be realistic and understand that the investment will be substantial and must be sustainable to provide for a comprehensive delivery system. The system and the way we can implement it and the preparedness level we have are only as good as their financial support, and if we do not support these efforts very much financially, then it is not going to be as proactive or good. Consequently, it will probably cost us more in the long run to do the recovery and the relief work.

So the financial part is a real key to the front-end load and the preparedness. Financial support is also a key element to effective recovery. We have dealt with many of those programs for years and years, not only on the public assistance side, but also on the individual assistance side and in mitigation. I think they are pretty well crafted because their specific parameters, criteria, and eligibility elements have been refined over the years and in many disaster responses. The recovery programs are not built to take everyone back to pre-disaster condition—that could not be—but it is the serious and necessary needs that are important. It is a shared responsibility as well, so does there need to be some tweaking? Absolutely—some tweaking to help us implement them better; some tweaking to help us decide when it should and should not be implemented; some tweaking in process, so there are similarities between program ability and the ability to network between those program providers so there is no duplication of benefit or overlap. Yes, some funding needs to be put into those mechanisms to enhance them, so we can lessen the duplication and make it more streamlined.

Of the seventeen different preparedness grants that are available, the ones we deal with the most are through the Department of Homeland Security. There is also funding that is put on the table at the state level, the local level, and the tribal level for preparedness efforts. That is also the case in the recovery side; there are many potential recovery programs, including, but not limited to, the Stafford Act, Farm Service Agency, Housing and Urban Development, and Natural Resources and Conservation. The key is matching the specific need to the appropriate resource, provided the thresholds for assistance are achieved.

Generally, implementing any of these programs requires some type of funding match or sponsorship. Every program has a different set of requirements. The tricky part is meeting those requirements and providing timely assistance. There is no one source you go to for funding, but each level of engagement needs to be appropriately funded so that the system can be effectively put together. An example of that is a mechanism comprising three parts, and all three parts are just as important—federal, state, and local levels.

On either the preparedness side or for response and recovery, because it is a system of all the components and parts, there are inter-relations and inter-reliance between those partners to do an effective job during disaster relief. If one of the pieces does not quite get what it needs, then that causes a gap in the system. That means that either someone else has to pick it up, or it will go unattended, and we just have to deal with the consequences of not being able to provide that. The more we can tighten that, the better off we will be, because then we do not have those gaps and those things we cannot attend to, which makes the recovery or response more difficult.

Changes

I think key critical foundational elements need to change. There needs to be a shift in culture from a compartmentalized development process to a collaborative teamwork process to develop the mechanisms and the procedures and the processes necessary to effectively deal with incidents, emergencies, and disasters. The development piece must be a grassroots effort knowing that all incidents, emergencies, and disasters happen at the local level. It is critical that all levels of service (local, state, federal) are integrated and coordinated. I think finding the common pieces, finding the disaster-related specific pieces that need to be tweaked, is important.

As an example, two very fundamental components (alert/notification and crisis management software) are not consistently implemented or coordinated across the country. We need to be able to alert and notify our citizens and the staff who respond. During events, we need to be able to communicate effectively and seamlessly within and between states to achieve a timely response. Finding those common operating components is extremely important toward the development of an integrated national system. Because of the level of complexity and different requirements imposed in each jurisdiction, it is understandable that some tweaking to achieve effectiveness will be necessary. The key is discovering and developing the most common national components and diversifying from that point. We need to look at it from a global approach, and all the cause-and-effect relationships must be taken into account to try to wrap our arms around this thing because it is very complex.

One of the things that make development of that system hard is that the magnitude and severity of what we deal with will never be the same every time we deal with it. It is always changing and it will always be different, so to put order to that complex adaptive system is a real task. I think the collaboration is an absolute key to putting some order to non-order. You can look at that in a disaster setting. When a disaster hits, it is all over the board, and you must put order to non-order and try to bring things together as close as you can to get the components to work effectively, to deal with what you have been handed. I think that is the key and the most critical piece.

Changes are happening across the board, not just at the local level, but at the state and federal levels, as well. I think we can do a much better job of developing what needs to be changed because many changes that occur several times are top-down-guided. A good example of that is NIMS (National Incident Management System) compliance, or NIMS implementation.

Driven from the top, NIMS is a federal Homeland Security directive, but if you actually look at the initiative, that is the kind of initiative that needs to be a grassroots initiative if we are going to have a system in place across the country. It needs to take into account all stakeholders and all levels of participation in the development, because everyone will have to use it and be cognizant of the different levels of involvement to put in place a system that will be effective. Many times we have gotten directives about what we will do, here is how it will go, and we say, "Stop, time out, it will not work that way." Some of those things occur more frequently than they should, where the collaboration on the front end is not as effective as it needs to be. Some of the research I have done shows that there is a real gap in true partnership and collaboration in development; it is traded for simple consultation with the partners. We see that daily.

NIMS is based on the Incident Command System. Developing such a mechanism has taken the Forest Service forty years to perfect, and they are still tweaking the system to provide continual improvement. The system is based on developing an operational capability to manage incidents with specific requirements for the various components. In the guidance for emergency management development, the NIMS initiative is treated like a check-off system of parameters that need to be met. There is not an understanding that the development is evolutionary in nature and needs to be strategically implemented. The impedance component rests with the fact that many of the players are volunteers, and the time commitment to implement the guided system is not practical.

The real kicker is that the compliance is tied to a jurisdiction's eligibility for grant funding. NIMS is about a process of developing operational abilities, not about compliance. Tying compliance to funding is absolutely ludicrous. The NIMS piece is just one example of where the collaboration is necessary, and the system development must be strategically guided to make sure we are really measuring the right things and really developing an operational component that is effective. It is one thing to train on the system, but the true test is actually the hands-on experience in developing the abilities of those who implement incident management to provide for an effective system. It is not about, "We are compliant this year because we did these things." You have to look at what the football is supposed to look like when we are done, and what is the strategy for getting there, and understand it will take a considerable amount of time and funding resources to develop an acceptable product.

Another one of the challenges we face is understanding that not all entities will move the football down the field at the same pace. Take a look at New York and the large urban areas, and then look at rural areas—different needs and situations. Should they be trying to capitalize on it? Absolutely. Will they develop them at the same pace? No, because they do not have the resources to do it at the same pace, whether it is financial resources or personnel.

The necessary changes I see in the response system are initiative development from a global perspective, being cognizant of the jurisdiction-specific needs, and providing for continual sustainability. You always have to be trying to improve and enhance the system. If you are looking at how we change the mechanisms there for providing relief, I do not know that I see any need for huge changes in the disaster relief picture.

I think what we need to do is always be cognizant of cultural differences, to keep them in the forefront of our minds. We cannot forget them; we have to keep them in the forefront because we have so many different nationalities in the United States. One reason I picked out the native nations is that we deal with seven native nations here in Montana, and even among those nations, there are some cultural differences and distinctions and expectations. If you were to look at all nationalities in this country, they all deserve the same fair, equitable treatment. We always need to keep those things at the forefront of our minds, because equity, fairness, and those cultural distinctions need to be part of the mix.

I think as we educate citizens more, involve citizens more, and empower the citizenry more to be a part of the solution, we will see improved involvement and preparedness. I think that is one of the keys to the effectiveness of relief, because the citizenry is part of the team. They happen to be the beneficiary of a greater share of the assistance, but they are still a part of the team that needs to come together to lessen the impact of the disaster and the amount of the assistance that is necessary. Each of us has a responsibility to do our best to be as prepared as we can be to mitigate the impact of an incident, emergency, or disaster. Preparedness starts at home. Effectiveness begins with the partnerships we develop along the way. Those partnerships will be the foundation for our successes as a nation in dealing with the events that test our patience, mechanisms, and operational capabilities.

Dan W. McGowan is a 51-year-old native of Montana. He was born and reared in Helena, Montana, located at the base of the Continental Divide in the heart of the Rocky Mountains. The mountains have always been home to him, and he continues to reside there.

In 1978, Mr. McGowan graduated from Carroll College with a Bachelor of Arts degree in communications and began his career in the corporate world. He traveled across the country as a district manager and joined the ranks of public employment in 1985.

Mr. McGowan has spent the last seventeen years in emergency management. The majority of his time has been occupied with the duties as the planning bureau chief for Montana Disaster and Emergency Services (DES). He accepted the additional duties as the agency individual assistance officer during disasters to effectively activate and deliver program assistance to individuals, businesses, and agricultural producers. In February 2004, he was promoted to the position of administrator for Montana Disaster and Emergency Services and is the Homeland Security adviser to the governor.

In April 2005, Mr. McGowan received the distinguished Small Business Administration Phoenix Award for his contribution to disaster recovery by a public official. He is currently involved in obtaining his Master's Degree in homeland security from the Naval Post Graduate School.

On a more colorful note, Mr. McGowan has a life outside of his commitment to emergency management. He volunteers for many community events, assists organizations with developmental issues, and is an accomplished rodeo announcer and a professional auctioneer.

A Critical Difference: Pre-Disaster Planning

Darrell Darnell

Director

District of Columbia Homeland Security and Emergency Management Agency

The Current State of Natural Disaster Relief

Emergency management agencies across the country have developed disaster response plans based largely on the concept of Emergency Support Functions (ESF). The ESF concept defines the roles and responsibilities of relevant agencies according to the incident and the requirements needed to respond to and mitigate the consequences of an incident. ESFs can be broadly defined as a grouping of government and certain private-sector capabilities into an organizational structure to provide the support, resources, program implementation, and services that are most likely to be needed to save lives, protect property and the environment, restore essential services and critical infrastructure, and help victims and communities return to normal, when possible, following an incident. The ESFs serve as the primary operational-level mechanism to provide assistance to state, local, and tribal governments or to federal departments and agencies conducting missions of primary federal responsibility. The emergency management agency for a local, state, or tribal government is responsible for managing this process.

Many emergency management practitioners subscribe to the notion that most individual citizens expect a competent, quick, and compassionate response to a crisis. To that end, agencies work to manage these expectations by doing a better job of educating and informing citizens regarding the capabilities of public safety agencies, and, as individual citizens, working with local governments to protect themselves and assist in relief efforts. Categorically, there is a general lack of understanding about the roles and responsibilities of public safety agencies (particularly at various levels of government) and some miscalculating what capabilities are in place to handle relief operations.

Public safety agencies such as emergency management, fire, police, health and human services, transportation, communications, procurement, state National Guard units, the Red Cross, and local and state-sponsored community/volunteer organizations are the primary agencies involved in natural disaster relief. However, depending on the severity of the incident, it is not inconceivable that all government entities would be involved in relief efforts. These agencies were designed and organized to provide support and services in times of emergencies and crisis and are generally organized along the ESF concept.

The financial components of natural disaster relief begin with a determination ahead of the disaster regarding what the financial impact of a disaster could be on an area. Planners develop comprehensive risk assessments to determine vulnerabilities and devise strategies to mitigate the consequences of a disaster; e.g., who will be affected, and what the costs will likely be for equipment, training, hardening facilities, etc., for preparing versus the costs of not preparing or providing relief.

Once a disaster occurs, the ESF structure largely defines the criteria for determining the financial implications and which agencies are involved in addressing the issues associated with the natural disaster. Additionally, under the National Incident Management System (NIMS), there is a Finance and Administration section, which has responsibility for coordinating the financial issues regarding an incident. NIMS provides a common and systematic, proactive approach for guiding departments and agencies at all levels of government, the private sector, and nongovernmental organizations to work seamlessly to prepare for, prevent, respond to, recover from, and mitigate the effects of incidents, regardless of size, location, or complexity, to reduce the loss of life, property, and harm to the environment. To receive federal funding from the Department of Homeland Security (DHS), all state and local jurisdictions are required to implement NIMS.

For the better part of the twentieth century, resource allocation for disasters was the responsibility of local and state governments, but the cold war and Civil Defense Act of 1950 granted the president of the United States authority to declare federal disasters and provide federal allocations. Based on the act, Congress can also determine fund allocation amounts. The Disaster Relief Acts of 1969 and 1974, along with the creation of FEMA in 1979, consolidated disaster relief efforts under one agency and created a formal process for defining a disaster and the process for allocating funds to local, state, and tribal governments, businesses, and individual citizens.

The Stafford Act was passed in 1988, and the Disaster Mitigation Act of 2000 established a national program for pre-disaster mitigation and provided funding to states that develop "Enhanced Mitigation Plans." The creation of the Department of Homeland Security further influenced the allocation process as FEMA was merged with DHS, and resource allocation focused on terrorism-related issues and less on natural disaster issues. After Hurricane Katrina in 2005, a number of changes have been made to the Stafford Act, and FEMA is still adjusting to these changes and implementing them. The most significant changes to the act focused on disaster relief operations clarifying the circumstances under which FEMA and other federal agencies would, and could, provide support and expediting the process of providing disaster relief funding.

It has taken some time for FEMA to implement the changes because of the natural tendency of large bureaucracies to adapt slowly to change, because the agency itself was undergoing reorganization, and because FEMA was in the process of revising the National Response Plan (NRP). The NRP provides a concept of operations for coordination among federal agencies in providing disaster relief support to local, state, and tribal governments. Formerly it was widely viewed as a cumbersome plan that did not effectively provide support to local emergency responders and was blamed for many of the government failures in response to Hurricane Katrina. FEMA has recently completed its revision of the NRP, and it is now called the National Response Framework (NRF). The NRF, according to FEMA, is a guide to how the nation conducts all-hazards response, from the smallest incident to a major catastrophe. It establishes a comprehensive, national, all-hazards approach to domestic incident response, and it describes special circumstances where the federal government exercises a larger role in disaster relief operations in which local resources are overwhelmed.

While the issues related to natural disaster relief are numerous, the overarching challenge of understanding the potential for a natural disaster and developing plans to mitigate the consequences of one is a major issue. Although it is nearly impossible to predict Mother Nature or the effects of a naturally occurring incident, it is possible to predict the eventuality of a naturally occurring event.

In the United States, we enjoy a vast array of climate variations and topographical differences. Jurisdictions rest on flood plains, in the path of hurricanes, on earthquake faults, and in the path of potential wildfires. To that end, emergency management regarding overall planning, response, and recovery is critical. How we will be able to respond to an event; determine what response capabilities we have or need to develop; develop training and exercises; test the validity of plans; educate the public on the potential hazards and how they can assist emergency responders in protecting themselves; and, finally, allocate sufficient resources to deal with a natural disaster, depends heavily on the sense of urgency placed on the need for planning inter- and intra-jurisdictionally. An element of the nature of emergency management is to either have or develop partnerships and agreements with neighboring jurisdictions that might be able to offer assistance in times of need. In cases of extreme emergency, when an incident may overwhelm available resources, jurisdictions have agreements with neighboring communities for mutual aid assistance in place.

Environmental shifts are making a tremendous impact on our ability to predict potential natural disasters. Greater understanding among planners of these shifts will have to occur for effective disaster relief plans to be developed. Additionally, the public's expectations of what government can achieve or how it should respond are also changing. Given the overall investment that has been made in support of emergency preparedness efforts since 9/11, public perception is that it is the government's responsibility to protect them against any type of hazard, regardless of the magnitude or type of event.

Legal Issues

As stated earlier, the Stafford Act and the Disaster Mitigation Act of 2000 are the two major laws that have a national impact on disaster relief operations. The Stafford Act provides guidance for FEMA and other federal agencies regarding federal support for disaster relief, and the Disaster Mitigation Act of 2000 assists local, state, and tribal governments in developing and preparing disaster mitigation plans. These laws also, in effect, provide guidance for governments and relief organizations in determination of their legal and fiduciary obligations, as well as some limitations of disaster relief support.

There really is not a template for which disaster relief situations will require legal assistance, but the ESF structure and NIMS provide a framework under which legal issues can be identified and adjudicated. Each has a legal function designed to provide advice and remedies surrounding disaster relief operations. Legal issues should be incorporated into pre-disaster planning so that any concerns are addressed and remedied prior to a disaster; however, the legal function assigned to the ESF and NIMS provides a nimble and quick method of addressing unanticipated issues that occur during a disaster. Attorneys involved in legal actions relating to natural disaster relief will require knowledge of local, state, and federal emergency management response planning procedures; executive orders that guide the planning; local, state, and federal emergency orders and powers; and local, state, and federal constitutional authority. Attorneys should also be prepared to provide individual legal assistance and legal assistance to small businesses that need help in navigating aid application requirements and are seeking assistance from agencies such as the Small Business Administration.

At the end of this chapter is a list of statutes and regulations associated with natural disaster relief.

The Roots of Change

The federal government has become more proactive in its response and is not waiting for the disaster to occur before engaging local and state officials in preparing for potential disasters, and responding rapidly to a "no-notice" disaster. In the past, the federal government would wait for a state to request assistance, but because of Hurricane Katrina and the delayed response of the federal government, that approach has changed.

Clearly, the response to Katrina was the major political situation driving the changes in disaster relief. The failure on the part of all levels of government to engage in coordinated and integrated planning and disaster relief efforts only magnified the disaster on the Gulf Coast. The economic cost of the cleanup and recovery efforts for Katrina will be felt for many years in many direct and indirect ways, and in the end will cost in the billions and billions of dollars.

According to the Congressional Budget Office, Hurricanes Katrina and Rita caused more economic damage than any other recent catastrophe in the United States. Estimates indicate the economic damage, for both insured and uninsured people, is at $140 billion. This economic reality is the driving force behind local and state governments to become more self-sufficient in preparing for disasters, and has caused the federal government to re-think its disaster relief methods and philosophy.

The government's response, at all levels, to the wildfires in California in the fall of 2007 provides an illustrative example of the changing nature of government intervention since Katrina. Emergency management agencies throughout California, led by the state's Office of Emergency Services, had engaged in significant coordinated pre-disaster planning, ensuring that response capabilities were adequate, and in cases in which resources were lacking, developed mutual aid agreements with surrounding local and inter-state agencies. The federal government, led by FEMA, did not wait for the state to request assistance, as had been the case in the past. FEMA began discussing federal assistance with state officials at the outset of the disaster and was prepared to immediately provide resources and assistance at the request of the governor. It is widely agreed that the coordinated planning on the part of local and state governments and the proactive approach of FEMA mitigated the consequences of the wildfires. Conversely, it is widely agreed that this level of pre-disaster planning and coordination was not present during the California wildfires of 2003, and its lack led to devastating consequences to some communities, particularly in Southern California.

Costs associated with changes in responding to and providing natural disaster relief generally vary based on the size and scope of the incident. However, in the mitigation planning process, an attempt is made to estimate costs required to prepare for and respond to an incident. Costs are generally categorized in the areas of facilities, infrastructure, equipment, personnel, training, administration, and supplies. Two reports may be useful in determining costs associated with natural disaster relief:

1. U.S. General Accounting Office, Disaster Assistance: Information on Federal Costs and Approaches for Reducing Them, Statement of Judy A. England-Joseph, Director, Housing and Community Development

Issues, Resources, Community and Economic Development Division, March 1998 (available online at www.gao.gov/archive/1998/rc98139t.pdf)

2. Congressional Budget Office, "Macroeconomic and Budgetary Effects of Hurricanes Katrina and Rita," CBO Testimony by Director Douglas Holtz-Eakin, October 6, 2005 (available online at www.cbo.gov/ftpdocs/66xx/doc6684/10-06-Hurricanes.pdf)

As previously noted, the federal government is changing to more effectively respond to the current issues and needs surrounding natural disaster relief. FEMA has reorganized to become more proactive and responsive to local and state needs. The revised NRF clearly defines FEMA as the lead federal agency for disaster relief operations and clearly states that the FEMA administrator is responsible for pre-disaster planning and that during disaster relief operations, the FEMA administrator reports directly to the President of the United States.

Local agencies are also changing their systems to become more self-reliant and engaging in regional cooperation with neighbors in emergency management planning and response efforts, rather than solely relying on the federal government. State and local agencies are engaging in much more robust pre-disaster planning activities, such as conducting vulnerability assessments, cross-discipline first responder training, and equipping and exercising of plans so they are ready to effectively respond to a disaster.

Organizations such as the National Emergency Managers Association have developed mutual aid systems, such as the Emergency Management Assistance Compact (EMAC). EMAC provides assistance when resources of a jurisdiction become depleted during a disaster. Neighboring states and localities have also entered into mutual aid agreements that define how they will provide support during a disaster. The challenge for emergency management agencies is to maintain these efforts so the adequate response capability is maintained and becomes sustainable. Budgetary constraints, other daily responsibilities and requirements of agencies, and changing priorities are just a few examples that contribute to the challenge of maintaining a robust pre-disaster planning program.

Because the Stafford Act is the major relevant authority for providing disaster management relief, it should be amended. Recent disasters have demonstrated that the act does not contain adequate provisions and authority to address disaster relief needs resulting from a catastrophic incident. The act also does not adequately address the threshold for defining emergencies and catastrophic emergencies. I would recommend adopting the key recommendations by New York University's The Center for Catastrophe Preparedness and Response. Adoption of these recommendations would reflect today's financial realities for providing disaster relief, and streamline the process and timeline for providing assistance to those who need it the most and incentives for businesses to rebuild. The recommendations may be found at www.nyu.edu/ccpr/pubs/Report_StaffordActReform_MitchellMoss_10.03.07.pdf.

Implementing Change

One begs to question whether the roles and responsibilities of local, state, and federal government are changing. Perhaps the roles and responsibilities are not changing. What is changing, however, is the recognition that planning and response efforts must be coordinated and integrated if disaster relief is going to be successful, although the coordination and integration continue to be a source of contention.

The majority of disagreements tend to center on roles and responsibilities of public safety agencies and each level of government (e.g., which agency is the lead agency, or incident commander, during a disaster or what the roles of the local, state, and federal governments are in making strategic decisions during a crisis). These issues are rooted in the different operational missions of agencies and the constitutional authority of local, state, and federal officials, which are often inconsistent with operational issues during a disaster. Many of these issues are being resolved through the creation of the state and local plans, the NRP, and NIMS. However, complete understanding and acceptance require continued coordination, planning, training, and education among emergency responders, appointed officials, and elected officials at all levels of government.

Local and state governments have strong motivation to coordinate improvements and changes between them—the requirement for states to develop multi-hazard mitigation plans to be eligible for federal assistance in the event of a disaster. These plans are developed at the state level and reviewed by FEMA, which allows officials at all levels of government to understand the potential hazards that may affect a geographical area, and the needs and capabilities of a state to effectively mitigate the consequences of a disaster. These plans are updated and reviewed every three years. The major difficulty with the plans is making sure they stay current, and coordination and integration of the planning effort is maintained among all levels of government.

Changes are being implemented at the local and state levels with more community involvement via programs such as Citizen Emergency Response Team (CERT) training and related citizen preparedness programs. These programs provide education and training to community volunteers to assist in disaster response and relief, resulting in better coordination and planning for disaster among public safety agencies at all levels of government.

A major challenge in all this is sustainability, particularly as it relates to citizen preparedness programs that focus on education and information. Often first-responder agencies fail to see the value of the "volunteer dollar" and overlook citizens' ability and willingness to contribute to response and recovery efforts. Further, depending on the amount of time that has passed since the last incident, there is a tendency for the public to "tune out" messages if they do not view a threat as imminent. There is an expectation that the government will protect against or be able to adequately respond to a disaster, regardless of the circumstances.

The Impact of Change

Planning, training, equipment, and resources have the greatest impact on changes in disaster relief. Recent disasters such as Hurricane Katrina, the California wildfires, and even smaller localized emergencies, all indicate the need for pre-disaster planning, training, equipping, and adequate resources that take a multi-discipline approach toward responding to disasters. The responses to Hurricane Katrina and the California wildfires indicate the impact attention, or lack thereof, to these areas has on emergency management agencies to provide an effective response to a disaster. Emergency managers intuitively understand the need to plan for natural disasters and have the capability to adequately respond to the most likely consequences of a disaster. However, budgeting adequately is always a challenge, as disaster preparedness efforts are in direct competition with other municipal and statewide priorities. Consequently, emergency management planners must identify minimum requirements based on likely scenarios, and do so regularly so that requirements remain relevant.

The creation of the Department of Homeland Security and the various reorganizations of FEMA have had a tremendous impact on disaster relief efforts. The creation of DHS led to the consolidation of many relief and security agencies under one organization, with the intent of allowing for greater coordination and integration of security and disaster relief functions and sharing of intelligence information. But it also resulted in confusion regarding the roles and responsibilities of the agencies that were folded into DHS, most notably the continued reorganization of FEMA and the creation of the National Response Plan and then the National Response Framework. The responses to Hurricanes Katrina and Rita are examples of how this confusion regarding roles and responsibilities led to an inadequate response at all levels of government.

Changes in pre-disaster planning have been developed and implemented so that citizens benefit from a compassionate relief and aid operation that quickly and effectively meets their needs. As we saw in the recent wildfires in California, the pre-disaster planning and response to the disaster, from all accounts, were very well thought out and well-managed. So the public benefited from a professional and competent response effort that was also compassionate and responsive to the needs and concerns of those affected by the disaster. The challenge in all this is not to become complacent, and to build on what was successfully accomplished, and improve in those areas in which the relief efforts could have been more successful.

Most groups, particularly those that deal with people with special needs and pets, are happy that emergency management agencies are placing greater emphasis on planning for the needs of those with disabilities and/or pets during a disaster. They also have reacted favorably to being included in the planning process in a much more meaningful way than in the past. Consequently, in the future first responders will be better prepared to assist people with special needs and those with pets.

The most important piece of information that emergency management professionals could share is that everyone—government, businesses, and private citizens—has a role to play in preparing for and responding to natural disasters. Government has to provide leadership and resources, and manage expectations; businesses have to work with government agencies and private citizens to develop programs and processes that will lead to continuity of operations after a disaster to mitigate the economic consequences of the incident; and private citizens need to be informed regarding the capabilities of relief agencies (public and private) and to be capable of protecting themselves until public safety agencies arrive to provide relief. If we can achieve this type of holistic and integrated approach to natural disaster relief, then our response to, and ability to mitigate the consequences of, a disaster will be much more effective, as we will be able to respond adequately to all types and scopes of disasters.

Cases, Statutes, and Regulations Relevant to Disaster Relief

The Robert T. Stafford Disaster Relief and Emergency Assistance Act: The principal legislation governing the federal response to disasters within the United States. The Stafford Act details how disasters are declared, types of assistance provided by the federal government, and cost-sharing arrangements among the federal, state, and local governments.

The DHS State Homeland Security Grant Program and the Emergency Management Performance Grant (EMPG): These two programs are the primary federal funding mechanisms that provide support to state, local, and tribal governments for disaster planning, training, equipment, exercises, and citizen education programs. More information on these programs can be found at www.dhs.gov

U.S. House of Representatives, "A Failure of Initiative: Final Report of the Select Bipartisan Committee to Investigate the Preparation for and Response to Hurricane Katrina": The Congressional investigation upon which this report is based brought increased attention to the governmental response to Hurricane Katrina and led to the reorganization of FEMA, significant changes in the Stafford Act, and revision of the National Response Plan.

The White House, "The Federal Response to Hurricane Katrina: Lessons Learned," a Report Submitted to President George W. Bush: This report, prepared by the Homeland Security Council, is the only authorized report by the Bush Administration detailing the federal response to Hurricane Katrina.

Congressional Budget Office, "Macroeconomic and Budgetary Effects of Hurricanes Katrina and Rita," Testimony by Director Douglas Holtz-Eakin, October 6, 2005: This report provided an immediate assessment of the short-term and long-term economic damage and consequences of Hurricanes Katrina and Rita and provided a basis for Congressional decisions regarding funding reconstruction efforts on the Gulf Coast.

"Demographic Effects of Natural Disasters: A Case Study of Hurricane Andrew," by Stanley K. Smith and Christopher McCarty, in Demography, *May 1996, pages 265-275:* This report detailed the effect of Hurricane Andrew on the population of Dade County, Florida. The report focused on damage to housing, insurance settlements, population redistribution, and geographic distribution of the population who did not return to Dade County. It provides a basis for estimating the demographic effects of a similar catastrophic incident and developing short- and long-term planning assumptions for rebuilding efforts.

The National Response Framework: The National Response Framework (NRF) is a guide to how the nation conducts all-hazards response—from the smallest incident to the largest catastrophe. It establishes a comprehensive, national, all-hazards approach to domestic incident response.

Homeland Security Presidential Directive 8 (HSPD-8): This directive provides the basis and direction for the NRF (above). It establishes policies to strengthen the preparedness of the United States to prevent and respond to threatened or actual domestic terrorist attacks, major disasters, and other emergencies by requiring a national domestic all-hazards preparedness goal, establishing mechanisms for improved delivery of federal preparedness assistance to state and local governments, and outlining actions to strengthen preparedness capabilities of Federal, State, and local entities.

National Incident Management System (NIMS): HSPD-5 provides the basis and direction for NIMS. NIMS provides a common, systematic, proactive approach guiding departments and agencies at all levels of government, the private sector, and nongovernmental organizations to work seamlessly to prepare for, prevent, respond to, recover from, and mitigate the effects of incidents, regardless of cause, size, location, or complexity, to reduce the loss of life, property, and harm to the environment. More information on NIMS can be found at www.fema.gov/emergency/nims/

Internal Revenue Service and the Small Business Administration: Each of these agencies has a host of guidelines, regulations, and laws that detail the type of assistance businesses may be eligible to receive as a part of natural disaster relief. The agencies also maintain information regarding individual state regulations and laws regarding natural disaster relief.

The Pets Evacuation and Transportation Standards Act of 2006, Public Law 109-308, 120 Stat.1725 (2006): This law mandates that state and local emergency management agencies develop policies and plans for sheltering and evacuation of pets during disasters.

On September 30, 2006, Congress modified the Insurrection Act as part of the 2007 Defense Authorization Bill. Section 1076 of the new law changes Sec. 333 of the "Insurrection Act" and widens the president's ability to deploy troops within the United States to enforce the laws. Under this act, the president may also deploy troops as a police force during a natural disaster, epidemic, serious public health emergency, terrorist attack, or other condition, when the president determines that the authorities of the state are incapable of maintaining public order. The bill also modified Sec. 334 of the Insurrection Act, giving the president authority to order the dispersal of either insurgents or "those obstructing the enforcement of the laws." The new law changed the name of the chapter from "Insurrection" to "Enforcement of the Laws to Restore Public Order."

This change to the Insurrection Act makes it much easier for the president to federalize state National Guard assets during a disaster. The National Governors Association has gone on record opposing this change, especially given that National Guard assets are stretched thin due to the deployment of significant troops for the wars in Iraq and Afghanistan. *Note:* As a part of the 2008 Defense Authorization Bill, Congress is considering repealing this provision, and it may be repealed by the time this book is published.

Darrell L. Darnell was appointed on March 19, 2007, by Mayor Adrian M. Fenty to serve as director of the District of Columbia Homeland Security and Emergency Management Agency (HSEMA). In that capacity, he is responsible for carrying out the agency's mission to reduce the loss of life and property and protect citizens and institutions from all hazards by operating and maintaining a comprehensive all-hazard, community-based, state-of-the-art emergency management infrastructure, including coordination and management of the district's response to emergencies and disasters of all kinds, both natural and man-made.

Mr. Darnell joined HSEMA after serving as director of the Urban Areas and Exercise Program at IEM, a Louisiana-based national disaster and homeland security consulting company. In that role, he managed IEM projects that helped urban areas and large cities improve efforts to prevent, respond to, and recover from terrorism and other disasters. He also was responsible for IEM's homeland security exercise projects for the U.S. Department of Homeland Security (DHS) and other federal agencies, as well as state and local homeland security and emergency management agencies nationwide.

Prior to joining IEM, Mr. Darnell spent eight years in the federal government, first at the U.S. Department of Justice and then at DHS. At DHS, he served as director of the Preparedness Division for the DHS Headquarters Operational Integration Staff (I-STAFF), where he oversaw DHS and interagency operational planning and execution of national exercise, evaluation, and preparedness assessment programs. He also worked with the Office for State and Local Government Coordination and Preparedness (OSLGCP) in the planning and implementation of national preparedness programs in partnership with state and local governments throughout the nation.

Before his assignment to the I-STAFF, Mr. Darnell served as the Local Programs Division director within the State and Local Program Management Division at OSLGCP, where he oversaw the development and implementation of anti-terrorism and counter-terrorism preparedness programs within the nation's local urban areas. Mr. Darnell was also a lead training program specialist with OSLGCP, where he helped state and local emergency responders and public safety officials develop training and technical assistance programs to address weapons of mass destruction response planning and emergency preparedness.

Prior to his assignment to the OSLGCP, he was a special assistant to the director in the Office of Community Oriented Policing Services (COPS) and a grant monitoring specialist, both at the Department of Justice.

Mr. Darnell has served as an adjunct professor in the School of Business and Management at Prince George's Community College in Largo, Maryland, and at the University of Maryland, University College-Asian Division.

Before beginning his career with the Department of Justice in 1997, he served honorably as a member of the United States Air Force (USAF), retiring November 1, 1997.

Mr. Darnell was a 2006 senior fellow at the George Washington University Homeland Security Policy Institute and is a recipient of the Founders Award from the Naval Postgraduate School's Center for Homeland Defense and Security, given for his role in the development of the center's homeland security post-graduate and executive education programs.

Dedication: *I would like to thank Millicent Williams, executive director of Serve DC, and Donneshia Taylor, special assistant in the District of Columbia Homeland Security and Emergency Management Agency, for their assistance.*

Education, Preparation, and Cross-Training in Disaster Relief

Henry Renteria

Director

California Governor's Office of Emergency Services

From the Pacific Ocean to the Sierra Nevada mountain range, from the deserts to its many waterways, California's sought-after landscape is the result of millions of years of natural disasters. The state's twisting, buckling tectonic plates that formed hills and mountains let us know they are still alive through mild temblors nearly every day. Pounding waves crash against the state's shoreline tirelessly; storms and wind knock down man's creations regularly; blistering heat and freezing temperatures take their toll on all living creatures and create weather emergencies. Yet the unique topography created by these extreme events is what makes California one of the most desirable world locations in which to live and visit.

All disasters are local. Therefore, all emergency management activities, including response coordination to combat disasters, start at the local level. People sometimes have the impression that if a major event happens, it is the federal government's responsibility to go in, take charge, and provide resources. Although the federal government may provide assistance, the first responders and incident command are the responsibility of the local jurisdiction.

When it comes to emergency response, all government levels are partners, and each has a specific role. California experiences such a variety and number of events that it developed the Standardized Emergency Management System (SEMS) in 1991 that is based on the long-standing fire incident command system.

SEMS Organizational Levels and Functions

Five organizational levels:

5 State

4 Regional

3 Operational Area

2 Local

1 Field

State - Statewide resource coordination integrated with federal agencies.

Regional - Manages and coordinates information and resources among operational areas.

Operational Area - Manages and coordinates all local governments within the geographic boundary of a county.

Local - County, city or special district.

Field - On-scene responders.

Five functions:

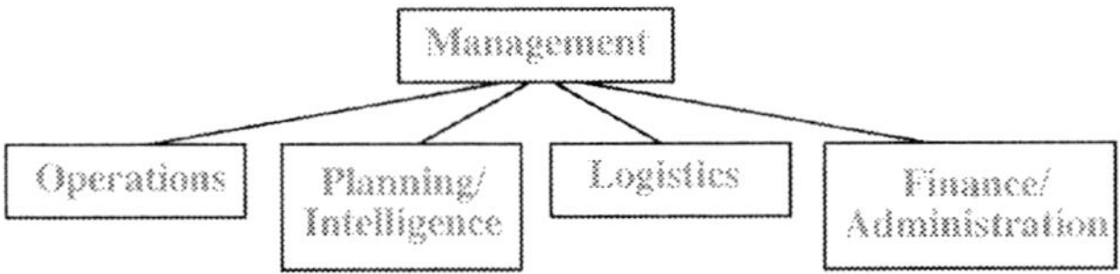

Management- provides the overall direction and sets priorities for an emergency.

Operations- implements priorities established by management.

Planning/Intelligence- gathers and assesses information.

Logistics- obtains the resources to support the operations.

Finance/Administration- tracks all costs related to the operations.

SEMS operates on the principle that since disasters are local, the first responders will be commanders and chiefs at the city or county level who manage their tactical resources and tactical event response. If, and when, resources are depleted, or close to being depleted, the SEMS provides a mechanism to request assistance from neighboring jurisdictions through mutual aid—a system of neighbor-helping-neighbor. If neighboring resources are insufficient to battle the event, resources from a "sister" jurisdiction are requested. California is divided into six mutual aid regional jurisdictions.

Under the SEMS, when state assistance is required, the state ramps up to coordinate response resources through one of its three administrative regions. OES regions include Inland, Coastal, and Southern, and provide direct support to an operational area. When a region needs support, it looks to the State Operations Center, located at OES headquarters in Sacramento, for assistance. Usually, it is only when all state resources and assets are inadequate to control the magnitude of an emergency that a request for assistance from neighboring states is made under the National Emergency Management Assistance Compact, or the state can request direct federal assistance under the Robert T. Stafford Disaster Relief and Emergency Assistance Act. This graduated system of broadening assistance levels has proved successful through many of California's disasters—most recently, the October 2007 Southern California wild land fires.

California's focus is to plan for the consequences of any event, whether human-caused or caused by Mother Nature. The state has experienced devastating wild land-urban fires in large, populated areas. The wild land-urban interface poses the unique challenge of fighting a forest fire in a densely populated area. Earthquakes seem to be more common in this state than anywhere else in our country, and in California, topography increases the challenge of hazard response.

When a large-scale emergency occurs in any type of disaster, planning is paramount to mitigating casualties, property loss, evacuation issues, medical surge issues, and sheltering. The disaster response requirements are what drive the state's emergency drills and exercises. Detailed information about California's emergency plan is available on the OES Web site at: www.oes.ca.gov.

Post-Hurricane Katrina discussions have also resulted in California's awareness and consideration of the need to enhance planning efforts to include catastrophic disasters. However, the term "catastrophic" has been difficult to define. California has the experience and resources, for instance, to tackle major wild land fires or devastating earthquakes, such as the 1989 Loma Prieta quake in California's San Francisco Bay Area, and the 1994 Northridge earthquake in Southern California. Yet events of the same magnitude might be considered catastrophic in other parts of the country or the world. So determining what is catastrophic is not a one-size-fits-all proposition.

Key Challenges

The first challenge that surfaces during events is the critical component of alerting, warning, and notifying the population that is at risk. The United States does not yet have a comprehensive alerting and warning system, although the U.S. Congress is currently addressing this issue. But we do have pieces of it, such as the Emergency Alert System (EAS) that transitioned from the old Emergency Broadcast System, used by the federal government to warn states of a possible attack from a foreign source. The EAS has to be supported by various local systems to alert and warn populations in specific areas.

There are other challenges with respect to systems that are used by local jurisdictions. For example, will siren systems or telephone systems be used, or will a public address system in a first responder vehicle be used to alert people? A significant challenge is the creation of an open and seamless communication system that enables messages to easily flow from local to state to federal sources and back to the local jurisdictions.

The second challenge is the ability to rapidly evacuate a large population—cities of 100,000 or more. The systems we have in place require a tremendous volume of resources to notify people and to subsequently move them out of harm's way. This includes special consideration for disabled populations, which is being addressed with a new component of the Governor's Office of Emergency Services, called the Office of Access and Functional Needs.

Once people are moved, they must be sheltered and cared for. To manage this type of ongoing challenge, California continually seeks new partners with resources, including the private sector and non-profit organizations, and we are working diligently to incorporate them into the existing SEMS structure. An example of a public-private partnership might be partnering with a major package carrier that has a fleet of vehicles that could be used to transport wounded people when the usual emergency resources have been exhausted. California has made strides in this type of partnering through legislation and Governor Schwarzenegger's Executive Order (S-04-06) for the Public and Private Partnerships.

During the initial days after a disaster, private-sector resources may be critical in augmenting the state's first-responders' resources and aiding California citizens and businesses. Recognizing this critical need, Governor Schwarzenegger signed Senate Bill (SB) 546 in September 2005 to help expand public-private partnerships and allow greater participation by the private sector in governmental emergency management efforts.

SB 546 (Government Code Section 8588.1) encourages formal relationships between government and the private sector to monitor the status of the important resources that are controlled by the private sector—from food to telecommunications equipment— during disasters.

A successful example of a public-private partnership is the effort launched in April 2005, the "Be Smart. Be Responsible. Be Prepared. Get Ready!" campaign. California First Lady Maria Shriver and OES, along with other partners, continue to encourage all Californians to be prepared for disasters by using steps described in the "10 Ways" brochure and other preparedness tips, available on the OES Web site. It is intended that this campaign will be expanded to incorporate business preparedness needs, as well as a focus on special needs populations.

Issues

Education and Preparation

Educating California's current population of 39 million on emergency preparedness is an ongoing challenge. Different world cultures include disaster preparedness in their day-to-day thinking. For instance, the Japanese take disaster planning quite seriously and include information in their school curriculum. Young children learn that they live in disaster-prone areas and practice actions to take to minimize the impact of an event. The public education system is a valuable tool.

If you live in California, or are one of the 500,000 people who move to this state each year, you should be aware that you may experience an earthquake or a flood. However, many new state residents are unaware of and unprepared for the variety of emergencies that can occur here. It is especially difficult to reach and educate those who speak a different language or have different cultures or customs. One of the key programs that First Lady Maria Shriver has established is the California Volunteers program, a strategic plan for community preparedness, details of which can be found at www.californiavolunteers.org/.

Citizens have come to expect that their government will take care of them during a disaster. That is our role; however, we cannot do it alone. When people prepare themselves, their properties, families, friends, and neighbors—and identify the types of hazards they may live in or work with—we will able to work together to minimize the impact of an event. Citizens should anticipate their needs and prepare to sustain themselves for seventy-two hours, for example, by keeping basic emergency supplies on hand, having evacuation plans, and having communication plans for staying in touch with family members if they become separated in a disaster.

Community Automated Notification systems are an emergency communications tool that can be used by city or county government to notify its citizenry. This tool uses combined databases and Geographic Information System mapping technologies to deliver emergency notifications to the community via telephone numbers that are included as part of an existing system, such as the local jurisdiction's 911 emergency system. It can quickly target a precise geographic area or be used to notify an entire area that will receive an emergency message.

Planning and the Nature of Disasters

What we know about disasters is that no two disasters are alike. Planning for disasters is a work in progress. Each has unique properties, issues, and populations that are affected. It is not practical to write a plan to respond to an earthquake or flood; however, a plan can be written to respond to the *consequences* that those events cause.

In our post-9/11 and post-Hurricane Katrina world, we learned about two very distinct events—one human-caused, the other caused by nature. Practice scenarios could not have outlined either of these events. Yet because of each crisis, those of us in this business are able to make major revisions to our planning and preparedness processes.

Funding

Our taxes are used to fund response agencies, such as fire, emergency medical, and law enforcement services. And when a disaster strikes a local community that needs financial assistance, our state and federal government may provide funding to help offset response costs and/or infrastructure that is damaged or destroyed by the event.

Here's the gap: there is not a steady stream of revenue or funding that contributes to disaster *preparedness.* In California, federal government funding for homeland security and terrorism preparedness is about $8 per capita. The money that is designated for earthquakes, fires, floods, and all other natural disasters is funded by the federal government at about $ 0.36 per capita. Discussions are under way to consider funding all-hazards preparedness, which would begin to close that gap. This would be a step in the right direction.

Legal Smorgasbord

Federal laws, such as the Stafford Act, were originally established for preparedness and recovery from devastation and infrastructure damage due to war. The Stafford Act was signed into law November 23, 1988, and has since been amended several times. This act constitutes the statutory authority for most federal disaster response activities, especially as they pertain to the Federal Emergency Management Agency (FEMA) and FEMA programs. However, assumptions that are part of this act do not address many of the types of catastrophic events that we deal with today.

Complementary laws exist in California, specifically, the California Emergency Services Act, which outlines the function and operations of the Governor's Office of Emergency Services, and the California Disaster Assistance Act, which defines implementation of specific OES programs. Details of these laws are available on the OES Web site, www.oes.ca.gov. Each state in the United Sates has comparable law.

At the local level, there are ordinances and regulations that cities and counties have established to address emergency issues in their jurisdictions. For instance, some cities have strict laws that require earthquake retrofitting before a building can be sold and title assumed by a new owner. This is an example of a great public policy that strengthens the community structures as they are transferred from owner to owner. Yet, this type of law is not consistent throughout the state; there is not a single overriding set of comprehensive laws because local jurisdictions have the right to govern at the local level.

Unfortunately, local jurisdictions cannot anticipate every major emergency that might occur. For instance, if a community lies within a known earthquake area, it might experience temblors ranging from 3.0 to 8.0 on the Richter scale. But does that mean *all* of the community's buildings should be retrofitted? Age and historical significance can affect the approach to altering a structure or simply tearing it down.

During the 1989 Loma Prieta earthquake, many heritage buildings in Oakland, California, were damaged but did not collapse. Repairing them was a very expensive proposition. Many of California's original buildings were constructed in the 1800s, are still used in many jurisdictions today, and have not been seismically strengthened. If struck by a major earthquake, some smaller jurisdictions would be hard-pressed to recover and rebuild.

The 38,000-Foot View

During a recent presentation to an emergency services association, I learned that many people in the audience were new to this business. This creates another challenge. How do you prepare someone for this type of work? Being an emergency manager means understanding the political system in your jurisdiction. It means understanding your first responder arena—fire, law enforcement, and emergency medical services, as well as social services, mental health, and crisis intervention components during and post-disaster. It requires knowledge of the economy and the economics of recovering from a major event. It also requires knowledge of how government agencies work together and the ability to coordinate the activities of different agencies.

My key message was that emergency management fails when coordination does not occur. Our role as emergency managers is not to *be* first responders; our role is to *coordinate* the agencies that are responding and rescuing people. Our role is to *coordinate* the agencies that will help rebuild the community and, hopefully, return it to a state of normalcy.

The reality of disaster response it that first responders are also the first to return to normal operations. Those who continue working on the disaster—maybe for years on end—are the social services agencies, building and public works departments, general services, personnel—all of the other departments that in a day-to-day operation would never consider the need for disaster preparedness. So, as an emergency manager, it is imperative to understand this dynamic and train representatives of those organizations that will continue to address a disaster long after its initial impact has ended.

These groups are a critical mass that needs to understand what their roles will be before, during, and after an event.

California has developed a robust training system to assist all responders who will be faced with the state's ever-evolving set of challenges. The California Specialized Training Institute (CSTI) began more than thirty years ago as a state military training organization that focused on managing civil disorder. Presently, CSTI's curriculum has grown to include a full spectrum of emergency management and hazardous materials courses. New and updated courses are added regularly. The institute has attracted participation from across the nation and world.

On the Horizon

Significant change may be forthcoming in regard to homeland security and terrorism preparedness. Much work has been done in recent years to prepare jurisdictions for terrorist events. This includes the purchase of bomb detection and chemical response equipment. But the further away we sense a threat, the less prepared we become. We need to resist this potential for complacency and continually identify our vulnerabilities. The threat does not change, which is why emphasis on *all-hazards* preparedness is the wise choice for the future.

Deteriorating infrastructure is also an issue that cannot be ignored, especially in large cities. Crumbling and old structures, stressed and worn water, gas, and electrical delivery systems, and many roadways and bridges could all use upgrading, repair, or retrofitting. The integrity of our infrastructure will be tested when—not if—a major natural disaster strikes. The breach of New Orleans' levees could happen in Sacramento and other parts of our state. Thanks to our current administration, levees in the Sacramento region are being retrofitted. However, an infrastructure weakness can probably be identified in just about any community.

California has an active disaster mitigation plan to address infrastructure. The state has undertaken significant mitigation planning efforts for three primary hazards: earthquakes, floods, and wildfires. Hazard mitigation is defined as "any action taken to reduce or eliminate the long-term risk to life and property from natural, human-caused, and technological hazards."

Examples of hazard mitigation actions include projects, such as home elevations, floodwalls, vegetation management, and seismic retrofits. Hazard mitigation is an excellent investment in state and local capacity building. Effective mitigation saves lives and property losses and avoids business and service disruptions caused by disasters. Findings have shown that for every $1 invested in mitigation, $4 of disaster losses is reduced.

Cross-training among agencies and emergency managers is also critical. In a recent National Emergency Managers Association survey, which includes all fifty state directors, results indicated that only six or seven state directors have ever managed a federal disaster that carried a presidential disaster declaration. Developing a mechanism to mentor new people who are coming into this field, cross-training current professionals, and archiving our experiences into an available worldwide database of shared knowledge is paramount to strengthening our emergency warning, response, and recovery systems.

In May 2004, Governor Arnold Schwarzenegger appointed Henry Renteria the director of the California Governor's Office of Emergency Services (OES).

Mr. Renteria began his career in emergency management in 1984 after a successful career in social services in Texas, where he was a trainer for the Crisis Hotline of Houston and spent ten years as the executive director for Crisis Intervention of Houston.

After relocating to California and briefly working as the emergency services coordinator for Contra Costa County, Mr. Renteria became the first director of the Oakland Office of Emergency Services in 1985. In this role, he participated in the Bay Area Regional Earthquake Preparedness Project. Following the 1985 earthquake in Mexico City, Mr. Renteria traveled to the region to conduct research to help prepare the Bay Area for a similar disaster. Because of his research, the Oakland Office of Emergency Services expanded, adding emergency services coordinators and further developing its preparedness plans.

Under Mr. Renteria's leadership, the Oakland Office of Emergency Services responded to numerous federally declared disasters that hit the region including the winter storms of 1986 and 1987, the Loma Prieta earthquake in 1989, the freeze in 1990, the Oakland fire storm in 1991, and flooding in 1996, 1998, and 1999.

During his tenure in Oakland, Mr. Renteria helped design and implement the original Emergency Managers Mutual Aid Program deployed for use following the 1994 Northridge Earthquake and was involved in the original design and implementation of the Standardized Emergency Management System (SEMS). He was also responsible for citywide disaster planning, response, recovery, and mitigation programs and management of the homeland security and hazardous material regulatory programs.

In 1995, on the one-year anniversary of the Northridge earthquake while Mr. Renteria was attending an earthquake conference in Kobe, Japan, a powerful temblor hit there. The quake devastated much of the city, killing nearly 6,000 people. Mr. Renteria and those in his delegation were unharmed and remained in Kobe for ten days, advising the Japanese government and further gathering information on earthquake preparedness and response.

Since his appointment as OES director, Mr. Renteria has directed response and recovery operations to thirty-five emergencies, including seven presidentially declared disasters. The disasters included the 2007 Southern California fires and the Angora fire in Lake Tahoe, the 2007 severe freeze, numerous winter storm emergencies in 2005 and 2006, and the 2004 Jones Tract Levee break in San Joaquin County. Additionally, Mr. Renteria has sought and received FEMA Fire Management Assistance Grants for fifty California wildfires.

Under Renteria's leadership, OES launched the state's first-ever Public/Private Partnership Initiative to support emergency planning and response and recovery operations. Mr. Renteria also oversaw the state's integration of its landmark SEMS system into the National Incident Management System, serves as chair of Governor Schwarzenegger's Governor's Emergency Operations Executive Council, and represents the Governor as chair of the State Emergency Council.

Mr. Renteria is a member of the National Emergency Management Association. He is past president of the California Emergency Services Association, a member of the International Association of Emergency Managers, and on the board of directors of the American Red Cross.

Developing Responses for Evolving Issues

Jimmy Gianato

Director

West Virginia Division of Homeland Security and Emergency Management

Current Issues Affecting Natural Disaster Relief

Disaster relief in America has taken on an entirely new meaning since September 11, 2001, and even more recently with Hurricanes Katrina and Rita. The power of broadcasting enables events to be viewed live on TV across the world, which has made the general public aware of events literally as they happen. America now watches as events unfold live. Many will never forget seeing the second plane strike the World Trade Center, the rescue of the Quecreek miners in Pennsylvania, or the Sago tragedy in West Virginia. We will certainly always remember seeing bodies floating in the streets of New Orleans. A more acute awareness of events and how we respond to them has raised the expectations of the American public.

In addition to new issues for disaster recovery, existing concerns still remain; these include the challenging question of how best to balance the responsibilities of individuals, institutions, governmental agencies, and non-governmental organizations in contributing to the recovery of individuals, families, businesses, governmental agencies, other organizations, and communities from the effects of natural disasters. We have become a nation where we expect the government to take care of our needs during a disaster and, in many cases, have forgotten that we need to be prepared individually to care of ourselves, especially during the initial hours of an event.

Individual aspects of these issues take center stage from time to time, but the basic problems remain the same. For example, should insurance be the primary resource for people rebuilding after a disaster? If so, then what is to be done about those who cannot or will not purchase adequate insurance? Do you ignore their needs because they did not secure adequate insurance? Additionally, if you do not ignore the needs of these people, then how do you provide this assistance without compromising your basic premise that insurance should be the primary resource for rebuilding after a disaster?

The answers to these types of questions form the basis for the laws, regulations, and policies that govern disaster recovery. Consequently, a primary issue becomes how to balance the expectations of the public with what government, insurance, the private sector, non-governmental organizations (NGOs), and others can do to help recover from a disaster.

Another key factor when considering disaster relief issues is the distinction between a disaster and a catastrophic event. Current relief programs were not designed to handle such large-scale events as Katrina and Rita; as a result, there were "on the fly" interpretations made of current laws and programs to try to make them fit. One example of this was to use public assistance under the Robert T. Stafford Act, which is designed to assist governmental agencies in recovering costs associated with emergency response and protective measures and to cover the cost of damage to public and certain non-profit agencies. It is also intended to assist in providing emergency and short-term housing solutions during the aftermath of disasters. The intention was to assist states and local governments in recovering from disasters that occurred within their boundaries.

During Hurricanes Katrina and Rita, all fifty states were asked to assist in the aftermath. Daily discussions resulted in the federal government attempting to make provisions of the Stafford Act fit the current situation. With evacuees from this event dispersed all over the United States, the current program with its normal processes did not fit. States were required to adjust their normal programs and processes to perform relief efforts for the evacuees that they would normally do for their citizens. Tracking of evacuees, processing of the individuals, reuniting family members all became issues. Systems were not available to verify residency; identification documents were lost; and the systems used to verify these documents in most cases were destroyed.

Special needs and populations are other important factors that affect disaster response today. The awareness of these issues became even greater with Katrina and serves as another example of the impact of the media to help bring awareness to important issues. Special needs populations include groups such as the physically and mentally handicapped. The emotional well being of victims is a critical aspect of recovery.

Yet another issue that compounds the complexity of disaster response is that of pets and animals. The traumatic impact to the wild animals and pets became vividly evident, as well, during Hurricane Katrina.

In the face of all these issues, few individuals or organizations have the ability to recover from a natural disaster using their own resources.

Therefore, it becomes a matter of what other resources will be available to help them recover. These can include insurance, assistance from NGOs, and assistance provided by the federal, state, or local governments. It may also mean there is no additional assistance, and the individual must bear the full brunt of his or her loss. An important role of government is to educate the public about what to expect during a natural disaster, to help them understand the need for self-support for a time, and to let them know what is available pre-disaster to assist them in the event of a natural or man-made disaster.

Previously, there was a lack of planning for catastrophic events at all levels of government. Even today, there is an ongoing debate as to what the definition of this type of incident is and who needs to be involved in the planning. First and foremost, if we are going to develop effective programs related to a catastrophic event, we must establish a thorough definition of what constitutes a catastrophic event. This is an issue the federal government is still struggling with.

There have been many attempts to define a catastrophic event, but varying jurisdictions interpret this in different ways. One definition that is being reviewed by Mid-Atlantic states is as follows:

> A catastrophic incident is defined as any natural or manmade incident, including terrorism, that results in extraordinary levels of mass casualties, damage, or disruption severely affecting the population, infrastructure, environment, economy, national morale, and/or government functions. A catastrophic incident could result in sustained regional or national impacts over a prolonged period of time; almost immediately exceeds resources normally available to State, local, tribal, and private-sector authorities in the impacted area; and significantly interrupts governmental operations and emergency services to such an extent that national security could be threatened.

Although this is not an official definition, it seems to fit most types of events.

Many plans include dealing with special needs and animals; however, these have not been as "hot" an issue as many others. Pubic awareness of events helps to bring these to light. This public awareness is best achieved through a public education campaign that includes all federal, state, and local advocacy organizations, both governmental and non-governmental. Public awareness campaigns, though well designed and well presented, are only as effective as the amount of effort placed on follow-up to them. There must also be some type of identification of these sectors to local and state emergency management officials so that adequate planning and resource management can take place ahead of and during emergency situations. Only through such public outreach ahead of time can these specific sectors of response operations be adequately served.

We must remember that there are many types of special needs. We tend to focus on the handicapped, but special needs may include hospitals, nursing homes, and prisons, as well as other types of facilities and individuals that may require some type of special assistance to take care of themselves.

The Evolution of the Issues

Issues are evolving because of a change in society's attitude. Currently, there appear to be two basic trends at work. The first is for the government to take an increasing role in disaster recovery. At one time, there was no standing federal disaster recovery program. If there was a disaster, then a one-time assistance package could be developed and applied to that specific incident. Later, in the early 1950s, a standing program was developed in an attempt to standardize an approach to providing federal assistance. Over the years, the programs increased the assistance available and reflected changing perspectives on disaster recovery. An example of this is the changing name and status of the agency charged with implementing these programs on the national level. The Federal Disaster Assistance Administration became the Federal Emergency Management Agency, Department of Homeland Security, or FEMA.

Running parallel to the trend toward increasing the governmental role in disaster recovery is the trend toward attempting to transfer the costs of recovery of certain types of disasters from the government to the private sector. The efforts to do so are made through the development programs, such as the National Flood Insurance Program, as well as through emphasis on mitigation and preparedness actions and the development of legislation, regulations, and policies that limit governmental assistance.

For example, individual assistance programs are based on the theory that the proper governmental role is to assist people to "get back on their feet," rather than to restore them to their pre-disaster situation. Thus, if an individual's goal is to be restored back to his or her pre-disaster situation, then the person must have non-governmental sources of funding, such as insurance or savings, or he or she must be willing and able to take on additional debt. As mentioned, we have witnessed the growing expectation of the public to provide this type of support; yet it does not exist with the programs that we have in place today. Congress is continually evaluating existing programs, examining the way those programs are administered, and revising those programs. In addition, more responsibility is being placed on the states themselves to provide for their citizens; further, many states have developed their own programs to meet the immediate needs of their citizens.

In some respects, we must realize that there will never be a perfect program or perfect implementation of a program for a disaster because, by definition, a disaster is a situation that cannot be remedied by a one-size-fits-all solution. This is why we must define "catastrophic," develop programs, and implement laws that deal with catastrophes as a separate category. This was a major lesson learned because of the response to Katrina. We must modify programs to fit the needs of the event, and laws cannot be so rigid that they do not provide the flexibility to the program administrators to use the common sense approach. The expected outcome is that we will achieve programs with enough flexibility to allow for common sense.

The tension between these trends remains a key factor in disaster recovery issues, and the shifting balance between the two trends depends greatly on the political and social climate at the time. For example, following a highly publicized disaster, such as Hurricane Katrina, there are usually attempts made to increase governmental support of recovery efforts. However, once the focus is on budgetary concerns, then the emphasis is placed on cutting the actual expenditures.

Ultimately, the most important component of disaster management is the ability to learn from events; after many disasters and catastrophes, most agencies review their plans to look at what can be done to improve response and to understand the lessons that can be learned from the situation. We must always learn from our mistakes and the mistakes of others. This was truly the case in the United States after Katrina, when the federal government instituted a full review of the response plans of all the states and an ongoing analysis to identify gaps in existing and proposed plans and resource levels. Ongoing education will be the key to improving and implementing any work done during disaster relief.

Agencies Involved in Natural Disaster Relief

There are many agencies used during disaster relief. Providing disaster relief usually takes two forms: the initial response to address the immediate needs of the citizens and the recovery aspect. Many states use different state and local agencies in their efforts. A prime factor to remember is that the initial response to all disasters is local. In many cases, local agencies can self-maintain for a short time, but their resources are quickly taxed. The agencies involved in the response and recovery may vary based on the type of disaster and must include government resources, as well as the private sector and NGOs, such as the Red Cross, Salvation Army, religious organizations, and citizen volunteers.

At the federal level, the agencies involved include: the Federal Emergency Management Agency (FEMA); the U.S. Department of Health and Human Services (USDHHS); the Department of Homeland Security (DHS), which includes many federal agencies; the Federal Bureau of Investigation (FBI); the U.S. Department of Agriculture (USDA); the U.S. Environmental Protection Agency (EPA); the U.S. Army Corps of Engineers(USACE); and, in some cases, Department of Defense (DOD) assets. Depending on the disaster, other very specific departments may be involved.

At the state level, examples of the agencies involved include: The National Guard, state law enforcement, and state-level health organizations. Another type of agency involved at the state level is human resources and its many subsets, including mental health, environmental protection, and agriculture. First response agencies are a core component of the response and recovery, since they are the ones on the ground both immediately and during clean up and recovery. Another key element of response is the administrative and logistic group. This group is responsible for the procurement of all supplies and keeping track of related expenses. All of these agencies operate under the National Incident Management System (NIMS) during a disaster.

Volunteer agencies play a vital role in response and recovery. These may include the American Red Cross, Salvation Army, religious and civic organizations, and citizen volunteers. In addition, the private sector plays an important role; companies such as Wal-Mart, Home Depot, Lowe's, and others have created disaster response units to assist in disaster response. The private sector has many of the needed supplies, as well as the ability to move these resources. We tend to forget in some case about the vast resources that the private sector has. We don't think of vendors such as Wal-Mart as a mechanism to move large quantities of disaster supplies; yet it has one of the largest fleets of trucks in the country.

Most of these agencies have statutory responsibility to respond during disasters. In addition, they have the key resources needed. It is human nature to assist others in need. Many of these responders are part of the community and want to do anything they can in restoring it to its pre-disaster condition.

The Roles of the Agencies

The roles of the many agencies are diverse and range from logistic support to direct support. All responses are local in nature, and the roles of the state and the federal government are in support of those local agencies. Since many disasters overwhelm local resources, these support roles may include providing direct support to backfill primary response, such as police, fire, and emergency medical services.

Other agencies will provide support based on their expertise. The National Guard in most states plays a key role in the response and clean up of a disaster. The National Guard can supply equipment and manpower and help restore communications capabilities in a very effective manner. States must also have plans to supplement their Guard units based on current deployments and other missions. With a main asset such as The National

Guard being stretched thin, many states and local agencies have resurrected plans to use other means to perform these critical functions. Such plans include contingency contracting for such missions as debris hauling, security, heavy rescue support, air support services, movement of assets, and other disaster response activities and responsibilities placed on The National Guard. In addition, EMAC, the Emergency Management Assistance Compact, must be utilized. During Katrina and Rita, EMAC missions were requested from most of the United States to provide assistance to Louisiana and Texas. These requests for assistance included National Guard units from other states to support local National Guard.

The state and federal governments both have a vital role in the recovery mission. Programs at both levels that have been designed to assist government, the private sector, and individuals must be implemented; this implementation is traditionally executed in a joint effort by FEMA and state emergency management.

Activities of these agencies vary greatly based on their expertise. Many of these agencies engage in ongoing planning and training to prepare for the next event. For example, in West Virginia, planning is ongoing and occurs in conjunction with many of these agencies to prepare for the routine disasters that we face on a daily basis, which include major floods, fires, ice storms, and chemical incidents.

In addition, West Virginia is playing a critical role in the planning for a catastrophic event in the National Capital Region. Such an event could lead to a large-scale evacuation of the population of the area. The ripple effects of such an evacuation would have major impacts on the transportation infrastructure and other public infrastructure, such as water and sewer, and require major support in caring for the needs of evacuees as they relocate to other areas of the country.

Political Factors that Affect Natural Disaster Relief

A number of political decisions and trends affect natural disaster relief. Recently, the scrutiny of the response by Congress and state legislatures to certain events has had a major impact. Congressional oversight has made federal agencies more responsive to the needs of the states and their citizens. Although there are a number of excellent public servants at all levels, bureaucracy hampers many from doing what they believe is right. The guidance in West Virginia from Governor Joe Manchin III has set forth a philosophy that it is our responsibility to minimize the human suffering that occurs, while using a common sense approach to taking care of our citizens.

The more major media attention that an incident receives that affects an individual politician's constituents, the more likely it will result in additional assistance; to the extent the plight of the victims receives little notice and/or the victims are not the constituents of the particular politician, the most influential factors then generally become fiscal. In simple terms, political leaders from affected areas or those areas that have been affected in the past are more likely to be more responsive to the needs of the individuals. Small events that don't merit national media attention are not as likely to result in significant changes to current laws and policies.

The impact of these political trends has been uncertainty. Those charged with administering programs develop understandings of how programs are to be implemented, and these understandings may govern their actions for years. Then, in a very short time, political decisions may completely change the legal basis of these programs or the interpretation of policies regarding them. Then, suddenly, the previously accepted procedures are no longer followed. This can lead to confusion and unfortunate, unforeseen consequences, such as the creation of ill-considered precedents.

Political leaders' lack of education relating to the level of assistance available can also create confusion for the public. This is why comprehensive training is essential. Training can include a host of topics, delivery methods, and audiences. As an example, outreach seminars for local jurisdiction leaders covering updates to federal and state disaster assistance programs are vital for these local leaders' understanding and level of expectation of higher levels of government and their responses. There are also distance-learning products that present the general roles and responsibilities of leaders at all levels for disaster and emergency response operations, ranging from policy decisions to the authorities they give to various emergency response officers in their jurisdictions.

Dependency on the federal government has led many states to believe that when a disaster occurs, the federal government (FEMA) will be there to handle the programs and provide the funding to take care of the events. This may not be the case. Emergency management agencies in the states are well aware of the programs, as well as the constantly changing guidance and interpretations of regulations.

Many factors contribute to confusion during a response and recovery effort. Although we tend to focus on the negative aspects of many of the programs, we must also remember that most of these programs have provided substantial assistance to local governments, the private sector, and individuals. I believe that elected officials understand the law to be malleable, and, when confronted with a situation they feel needs to be addressed to help people, they act to change the law to meet the situation at hand. Sometimes they act under pressure, but in most cases they act in what they believe is the best interest of the citizens. Given the stark scenes of loss of life and property presented by disasters, and the information provided by victims and special interest groups, the tendency is for elected officials to act quickly.

The conflict here is that those who implement the programs based on the laws that are passed must see them as fixed; they are therefore required to implement what exists in legislation through regulations that carry the force of law and policies that often reflect a great deal of experience in implementing an authorized program. They are also often mistrusted by politicians and the public.

Therefore, under the stress of trying to assist large numbers of people in great need, there is, in some cases, an overwhelming desire to change programs, enact special programs, or undertake other legislative steps based on non-governmental advice instead of following the proven course or making only changes advocated by those who implement those programs. Again, many of these programs fail to use the "common sense" test. It is easier to tell a bureaucrat to fit a round peg into a square hole than to tell those in need there is nothing we can do to address their needs as quickly or completely as they desire, given the present laws. Disaster recovery is no different than any other situation in that "the wheel that squeaks the loudest gets the grease," and obviously this does not always result in fair or consistent actions.

The use of the regular military in disaster response is a new concept, as is allowing the president to take control of a state's National Guard during a disaster. This highly controversial issue allows the president to potentially commandeer forces normally under the control of a governor for use in his or her own state. If the governors of the states lose control of their National Guard, then this is a major asset lost.

The Financial Components of Natural Disaster Relief

The financial components of disaster recovery include individual resources, insurance proceeds, donations from individuals and non-governmental organizations, loans from either private sources or government agencies, and grants or tax incentives from local, state, and federal governments. Most federal assistance during a disaster is on a cost-share basis with the state or local government. The question that surfaces in a struggling economy is always where the funding comes from to support the response and recovery.

Depending on the size of the event, those who are financially affected can range from one person to the entire nation. Furthermore, depending on the nature of the event, it is conceivable that whole communities' tax bases can be destroyed, never to recover.

The challenge is to determine in each case the amount of financial assistance needed by each person affected and the sources, if any, that contribute to meeting that need. If current sources are not adequate, then decisions must be made as to what, if any, new resources will be provided through governmental or non-governmental sources. It also involves a balance in long-range planning and mitigation programs. In some areas, buying multiple properties and relocating citizens can have a positive impact. In others, moving those individuals from an area and having them relocate outside a jurisdiction could have major financial implications to government, education, and the private sector.

All agencies, ranging from the federal government to local political sub-divisions, are involved in handling the financial issues, as are the private sector, NGOs, and individuals.

The Role that Attorneys Play in Natural Disaster Relief

The roles of attorneys in disaster relief are similar to their responsibilities every day. During disasters, many attorneys volunteer to assist individuals in replacing lost documents, such as deeds, wills, titles, and papers of that nature. Attorneys also have played roles in developing class action lawsuits related to damage and negligence, as well as defending individuals and others who have been the subject of litigation. Depending on what services an attorney would like to provide, the common legal skills are required. If involved in litigation, attorneys may need specialties in various aspects of federal law, including the Robert T. Stafford Disaster Relief and Emergency Assistance Act (Stafford Act) (Public Law 100-707).

With respect to attorney involvement and the services they can provide, there is really no difference between those skills required for disaster relief and other aspects or specialty fields of law. In some situations, laws and regulations are suspended, and legal interpretations for response agencies are necessary. Knowledge of contractual law is also needed in many cases.

Government Systems that Are Changing in Response to Current Natural Disaster Relief Issues

Government systems from the executive to the division level are changing based on the current political structure and have periodically placed a higher or lower importance on the arena of emergency management and its activities. After September 11, 2001, the federal government underwent a significant reorganization with respect to emergency response and intelligence gathering agencies. During the last seven years, this reorganization has translated into changes at the state and local levels, as well. Mandated requirements for training (NIMS) and reorganized interaction among federal, state, and local entities have caused some friction, as well as redundancy and disorganization; however, the attempt to combine and coordinate like entities into an ever-changing viable effort at the federal, state, and local levels, while not easy, has been necessary. The result is to try to provide assistance to governments and communities in a timely, efficient, and effective manner.

Paul Howard, director of operations, West Virginia Division of Homeland Security and Emergency Management, notes that the horizontal relationship between federal agencies and the relationships between DHS/FEMA and the states has changed in several ways. Primarily, the roles of some federal agencies have evolved over time.

Most government systems have changed in the past several years because of the major events that have unfolded. Federal changes have been significant to the lessons learned from 9/11 and Katrina. The use of the regular military, as mentioned above, and the ability to federalize troops under the Insurrection Act of 1807 (10 U.S.C. §331 – 10 U.S. C. §335) are major changes in the use of military forces during disaster response at the national level, providing additional resources across the country. Mandated requirements for training, exercises, and planning, as well as the implementation of the National Incident Management system, have resulted in significant changes. While these changes were made swiftly after 9/11, the results have included some positive and necessary changes, as well as many that have simply caused chaos and disconnect.

State emergency management agencies are also affected in this time of transition, with an increase in message traffic and an increase in the number of potential supporting agencies with whom pre-event coordination must be accomplished. The result is, hopefully, a better coordinated response effort with a wider range of potential resources being brought to bear in a timelier manner. With this in mind, all federal supporting agencies are being placed on a footing of quicker response operations for all situations.

The Emergency Management Assistance Compact (EMAC) is another system that stepped up and has significantly improved and evolved because of recent disasters. This state-to-state mutual aid system was implemented with great success during Katrina and Rita, and is a vital tool for emergency management. As the EMAC Web site states:

> EMAC, the Emergency Management Assistance Compact, is a congressionally ratified organization that provides form and structure to interstate mutual aid.
>
> Through EMAC, a disaster-affected state can request and receive assistance from other member states quickly and efficiently, resolving two key issues upfront: liability and reimbursement.[1]

The emergency management agencies are also significantly affected by greater roles and responsibilities, including regional planning and increased support to other states. Program implementation, such as NIMS, has also proved to be a logistics problem, with mandatory coordination and training at levels of government that were previously not required.

Factors Driving the Changes in Natural Disaster Relief

Disaster response and management, as well as the consequent recovery aspects, are constantly evolving. Even without changes in laws, regulations, and programs, we learn from every response and strive to make improvements. I believe the major changes we will see in the future will depend on how we respond in the present. Many states and the federal government are being more proactive to make sure that the capabilities are in place to render aid when necessary. We are looking more closely at special needs populations and dealing with pets and animals better. We are also looking at sheltering larger quantities of people and moving them further away. Pre-Katrina, this was not a major consideration; now, catastrophic evacuation planning is mandated by federal laws.

Another driving factor for change is the availability of worldwide instant communication; natural disasters are no longer separated by time, distance, and economic factors. Further, environmental changes have caused an impact. There is no longer a part of the world that is safe from the ravages of natural disasters. Volcanic eruptions occur in Krakatau and Washington State. Flood waters ravage towns in Bangladesh and Great Britain. Refugees wander the streets of Mogadishu and New Orleans. If a disaster occurs in the Middle East or Africa, the Western world will no longer wait for a

[1] EMAC Web site, www.emacweb.org.

newspaper or an electronic message days later to report the specifics. If it is a local disaster, the citizens demand immediate action. If it is a foreign disaster, action is demanded without a seventy-two-hour time period.

Disasters have a significant impact on the economy. As mentioned above, a whole community can be destroyed and its tax base wiped out. The political reality of how to deal with those situations can be devastating to a career.

New Orleans is the perfect example. With the city located below sea level, it is prone to flooding again and to major devastation. This prompts the question of whether it is prudent to spend taxpayer dollars to rebuild that city in that location. Doing so is contrary to the mitigation programs many states develop, whereby they acquire all property that is prone to repetitive loss and relocate the citizens. Will that policy work there? What is the impact on the economy of the region and the state if the city were not to be rebuilt?

The political implications would be enormous. Imagine the impact to the City of New Orleans if a decision was made not to rebuild any portion of the city that lies below or in a flood plain. The tax base of that area would be destroyed. The makeup of the city government would be affected. Taxes used to fund critical services, such as education and public safety, would be lost.

In southern West Virginia, after devastating floods in 2001 and 2002, many homes were destroyed. Mitigation funds supplied by FEMA allowed local jurisdictions to purchase property that had been significantly damaged or had been repeatedly flooded. The major impact on the local economy was that many citizens moved away from the area. Property that was on local tax roles now became the property of local governments and was no longer taxed. This resulted in a significant loss in revenue for the jurisdictions. Losses in enrollment were also caused by people leaving the area, since regulations prohibited rebuilding on the property, resulting in loss of other funding for education based on enrollment. Political careers have been lost or severely damaged by incorrect decisions during a disaster and failing to develop programs to properly recover from these events.

These changes are being implemented at every level of the government. At the state level, changes are implemented by modifying plans and procedures to reflect the changes in federal law, regulations, and policy that govern the programs.

The challenges of implementing changes to recovery programs are the same as those involved in implementing the programs in the first place. Are there adequate numbers of trained personnel and resources available to do the job? In a non-disaster situation, implementing a change may be as simple as modifying a plan section. Receiving the guidance from the federal government in a timely manner is also an issue.

The challenges are overcome by providing adequate trained personnel and resources to complete the required tasks. They are also overcome by developing good working relationships between state staff and their federal counterparts.

The expectation is that the programs will be implemented in accordance with the legal, regulatory, and policy requirements in place at that time. Once the guidance for the program is received, these changes to policy and practice are implemented when the program affected by those changes is next implemented. Prior to that time, changes to policy and practice are incorporated in plan modifications and reflected proactively in the training of personnel who will administer the program.

Many policies change from disaster to disaster. It is an evolving process based on the circumstances. This is an ongoing process. A couple of examples of recent changes are as follows:

- The cap on federal disaster housing repair assistance has been changed from approximately $6,000 to approximately $27,000.
- The rules governing administrative costs for administering the public assistance program have been changed from a sliding scale system to a defined percentage of the grant amounts, and the state must determine how to divide it between itself as grantee and other recipients as sub-grantees. This places greater burden on the states in a major event.

The Impact of Changes to Natural Disaster Relief on Citizens

The impact that natural disasters have on citizens varies widely. As an example, federal law has been changed so that the cap on federal grant assistance for the repair of a home (real property) has been lifted from the former total of approximately $7,000 per applicant to the total amount of the federal Individual and Households Program grant of approximately $27,000. This will definitely benefit those people trying to repair their homes. However, it will not benefit those people who previously received the total Individual and Households Program because of combined real and personal property losses. Since the total grant cap has not been changed, an individual with substantial personal property loss, if he or she would have received the maximum grant anyway, may receive proportionally less money for replacement of personal property because a proportionally larger portion of their grant will go for repair or replacement of real property.

In the same way, the recent decision to eliminate the use of recreational vehicles as a disaster housing resource will benefit those who would have been adversely affected by living in them. However, in many cases they were a very valuable resource that allowed people whose homes were located in the hundred-year floodplain to live next to their flood-damaged home while repairs were being made; the RVs could rapidly be moved in case of additional flooding. Without them, the damaged home's owner will have to accept housing assistance that may be located miles away from his or her damaged dwelling.

While the benefits vary on a case-by-case basis, the key challenge is consistent regardless of the circumstances, and that is to implement the programs, regardless of changes, in such a manner that all legal requirements are met and the maximum assistance possible is provided to those eligible to receive it.

The History of Resource Allocation in Natural Disaster Relief

Resource allocation is directly attributed to the size and scope of the disaster. Resources range from supplies and equipment to the funds to implement programs and match grant assistance. In most disasters, the allocation of resources has been effective, since it is rare that multiple states compete for the same resource; however, during recent events the resources were not available when needed. FEMA was not as prepared as it needed to be to get the resources where needed.

The primary factor influencing resource allocation in most situations is funding. Recently, FEMA and state emergency management agencies have been more aggressive, and Congress and state legislatures have been more receptive to providing funding for many of these programs.

I believe that the two main reasons that resource allocation is changing is the increased public awareness of need and the political reality that being unprepared is not a politically correct way of handling situations.

The basic changes to the area of natural disaster relief response and resource allocations include modifications to existing programs so that they reflect political priorities, concerns for health and safety, and a range of other concerns. The examples above illustrate these changes in some areas, and other areas include increased emphasis on the evacuation and sheltering of those with special needs and the care of pets in sheltering situations.

These changes are made through legislation and specifically by changes in regulations and policy modifications to improve our capabilities and responses. Further, we use past events as lessons to teach us how to better our future responses.

Despite these changes, many of the roles and responsibilities of those involved remain the same. The National Response Framework is the latest guide for how we respond to and recover from events; however, the long-term recovery continues to be an evolving issue, and in many cases is guided by the events themselves.

Most changes are made immediately upon release of the guidance from the federal government, thus requiring the states to modify their plans and programs and implement the adapted versions. In these instances, the challenge for the states becomes the ability to keep up with continually evolving program guidance and new programs. Many state staffing levels have remained the same while the workload has increased. Local governments look to the state as the authority on these programs and to provide them guidance. In many cases, the local agencies expect the state to complete the work on their behalf, which can lead to dashed expectations and delayed responses.

Mechanisms Used to Coordinate Improvements and Changes among Local, State, and Federal Agencies

Changes in disaster recovery programs come from the federal government directly to the states. In response, we at the state level make the required modifications in the programs that we administer, such as public assistance, and gain as much information as possible regarding changes to programs that we do not administer, such as disaster housing, so we understand the modifications and their implications.

That information must then be passed to local agencies so that they can develop their plans and understand the parameters under which they must operate. Similar procedures are used with other state agencies so that the information is disseminated. Other mechanisms used are various training programs developed by federal agencies, the state, and the private sector.

Important players not mentioned elsewhere in this document are the national organizations that represent key stakeholders in Washington; examples of such organizations include the National Emergency Management Association (NEMA) and the International Association of Emergency Managers (IAEM). They are constantly monitoring legislation and programs and passing information to the states; they also serve as advocates for the states at the national level.

Programs are delivered in a variety of formats, such as in-person meetings and training sessions, Webcasts, conference calls, and newsletters. A principal concern is to have the modifications in place before an actual implementation of the program must be executed. We have had instances in the past when program changes were being made on the fly, and this caused a great deal of confusion, wasted energy, and frustrated assistance recipients.

It is also challenging to find the time and resources to institute the programs and attend the necessary meetings. In addition, many of these require extensive and expensive travel for the states. It is difficult to develop a good solution for this situation. If staffing levels are not increased, the same number of people has to be spread over ever-increasing time and project requirements. The potential result is that important work and/or meetings are not accomplished at appropriate levels, and this benefits no one in any field associated with these programs. The governing question is what the most important thing is at the time, and that may not be the most important thing to do over the longer term.

Although these challenges are not insurmountable, they still raise issues for the states. Alternative ways and innovative thinking must create solutions to overcome the issues of meetings across the nation. With technology as it is today, some of these challenges can be met by video conferencing, Web conferencing, and conference calling. However, in many situations, face-to-face meetings are required. Even new technologies will allow participants in meetings to literally sit across the table from one another in virtual meetings. Cisco's new Telepresence solution creates a conference room environment in a virtual setting so real, it is hard for the participants to realize they are not in the same room.

There are no obstacles that cannot be overcome with the desire to do so. By bringing the appropriate people together, issues can be raised and solved, new programs vetted, issues managed, and interaction between peers accomplished.

Recommendations for Managing Natural Disaster Relief Programs and Efforts

The basic concept of natural disaster response and recovery is that all disasters are local, and that the response efforts must begin at that level. Subsequent actions by the state and federal governments are intended to supplement the efforts of the local government in providing for the safety and well-being of its citizens. As state and/or federal assistance programs are implemented, they are conducted in accordance with the relevant laws, regulations, and policies that govern those programs. Hence, in a sense, the local government loses control of some aspects of the recovery as these additional forms of aid are brought in. For example, if it is determined that a mobile home group site must be developed to provide housing for disaster victims, the local government, FEMA, and the state will work jointly to decide on its location. It will no longer be simply a local decision.

However, all of the state and federal recovery programs should be seen as tools that the local government and its citizens have to help them in their recovery from the disaster. Sometimes the tool does not meet the desires of those affected, but it then becomes their responsibility to match other resources with the tool to realize the fullest possible recovery from the event.

This process is somewhat customized for each situation. Each disaster has distinctive aspects, and the political and program personnel managing the recovery efforts and programs respond to these unique situations with as much consideration and flexibility as permitted by the laws, regulations, and policies governing the disaster recovery programs and procedures. The primary goal is to use the available resources to meet the actual needs to the greatest extent possible.

In all scenarios, ultimate responsibility for the recovery of individuals and families rests with those individuals and families. Final responsibility for the recovery of communities rests with the leadership of those communities. Ultimate responsibility for the governmental response and recovery efforts rests with the local, state, and federal political leadership. For example, on the state level, the governor will, in the case of a presidential major disaster or emergency declaration, appoint a state coordinating officer to manage the state's recovery efforts; however, ultimate responsibility for the state's efforts rests with the governor.

Adapting to new policies and procedures necessarily requires considerable time and money. At this time, the change that seems to be eliciting the most attention is the incorporation into law of the need to have designated special needs and pet shelters in addition to shelters for the general population. Changes to the recovery programs primarily require administrative adjustments at this time. Their actual implementation may result in increases in the amount of time and money needed to achieve their goals, but we will not know this for certain until we have direct experience with the modified programs.

Also, catastrophic planning will require a large amount of time and resources and may be necessary on only a very small percentage of events. In most cases where states face a major declaration under any new programs, there is typically a 25 percent cost-share to the states for most of the programs implemented.

Prior to the recent changes, the focus was on providing adequate shelter to all groups, but it was a goal that, if left unmet, did not mean that a law had been violated. Since the change in the law, this is not the case; as a result, all involved in sheltering and evacuation must dedicate more attention to the letter of the law, rather than to the basic goal of simply providing humanitarian assistance to those in need.

Katrina dramatically illustrated the need for the nation to have in place plans to move large numbers of people to a wide area. With Katrina, evacuees were moved to all fifty states. As exemplified by this tragedy, changes in legislation, procedures, and policies are brought about by major events. In addition to legislatively mandated changes, there are those related to health concerns. The recent decision to eliminate recreational vehicles as a means of disaster housing is an example of a change due to health concerns. This may make it more difficult to house people in rural portions of the states. However, we will not know the complete effects until FEMA has determined the alternatives to recreational vehicles and how they will be implemented.

Recommended Methods for Those Navigating the Changes to Natural Disaster Relief

The methods to be followed when attempting to navigate the changes to natural disaster relief are largely dependent on the knowledge a person has of the pre-change disaster relief policies and programs. The more knowledgeable one is of the past, the easier it is to note and adapt to the changes. If someone is trying to navigate the disaster response and recovery world for the first time, I would advise them to forget trying to concentrate on the changes and to concentrate instead on the current law, procedures, and policies. It is these policies, rather than those that existed in the past, that will govern the assistance provided and the manner in which it will be provided.

With respect to resources, I recommend beginning with the Web sites for governmental and non-governmental organizations involved in disaster response and recovery operations, and follow the various aspects of the programs and efforts from them. For example, with respect to sheltering and evacuation, one can access the federal laws, regulations, policies, and plans on the FEMA.gov site; then, it is effective to move to the plans of individual states and local governments, as well as to the information provided by other governmental agencies and NGOs, such as the American Red Cross. In addition, there are many sources one can use to access the U.S. Code and the various Codes of Federal Rules (CFRs).

The resources mentioned above become important to those navigating this process because the various sites and organizations spell out as clearly as possible the existing rules and procedures. Disaster response and recovery is an inherently vast and complicated area, with a number of governmental and non-governmental agencies involved; it also encompasses a wide range of documents, from laws to policy statements on individual aspects of the implementation of a program. In addition, interpretations of policies are sometimes made that, in clarifying a policy, may alter long-held ideas about the policy. It is challenging for those not immediately involved in implementing a program to be confident they have the latest, most up-to-date information, and, even if they do, it can change very quickly. Therefore, after a basic understanding of the processes and programs is gained, it is best to contact those implementing the programs to determine the current nuances that govern the process of providing the assistance authorized by the relevant laws.

Persistence is a key to success. It is important to remember that the volume of laws, programs, regulations, and policies will most likely be significantly larger than the areas with which one is directly concerned, and many of the items listed may be better provided via links and other references to enable users to concentrate on the most critical and basic elements of interest.

The Beliefs of Groups and Individuals Regarding Natural Disaster Relief: Misunderstandings and Mismanaged Expectations

Most groups and individuals believe that response programs should provide for their safety and well-being during and immediately after a disaster event.

They further tend to believe that recovery programs should undo the effects of the natural disaster by returning them to their pre-disaster situation or, perhaps, even improving their lot. In some cases, such as individuals with adequate insurance coverage, this may actually be the case, but too often, it is not. Many individuals depend so heavily on the government to look after them that they are unprepared to help themselves in the initial hours of a disaster.

Expectations in many cases cannot be managed by the agencies involved, but we attempt to manage these expectations to some extent by being forthcoming as to what the various programs can and cannot do. We communicate this information through press releases, media appearances, and contacts with various officials and affected individuals. For example, we emphasize the need for individuals to develop their own family emergency plans and supplies so they can manage on their own for a period of days following an incident. In addition, we promote the purchase of insurance to increase the likelihood that an individual, family, business, or other entity will have the financial ability to recover from the disaster.

Two major misunderstandings are that the goal of disaster recovery programs is to restore individuals and families to their pre-disaster situations and that disaster assistance for publicly owned property is to solve pre-existing non-disaster related problems of that property. Many people are frustrated to learn that recovery programs will not replace their Christmas decorations or swimming pool, and many local officials cannot understand why disaster funding will not replace a totally inadequate water distribution system, but only that portion damaged by the disaster.

Another key misconception is the idea that government will always be there to provide for each individual. In some cases, the government agencies that normally respond may be as affected as the individuals requiring assistance. An example of this is a situation such as a pandemic, wherein the agency itself may suffer limitations or other impairment.

These realities are key issues that must be understood by anyone potentially affected by these situations. Everyone must understand that nearly all disaster assistance funding has procedures, policies, and limitations that must be followed by those administering the programs. Even when there are goods or money donated through voluntary agencies, those agencies generally establish guidelines they follow in distributing those resources. There may be rare instances when there are no limitations, but these are few and far between.

Agency Factors that Have the Greatest Impact on Natural Disaster Relief

The major factors that have provoked changes in agencies are leadership, training, resources, and experience.

Leadership in an agency can assume a greater or lesser role in modifying existing or seeking new means of providing disaster assistance. Training will provide the knowledge of existing programs and a better ability to implement them. Resources, or the lack thereof, determine what can actually be done and the time frame in which it can be achieved. Experience provides a guide to the most efficient and effective response and recovery techniques based on past successes and failures. The changes these factors can bring about are within limits established by law and regulations, and must be in line with the policies of the governor. A major issue in many states is that the leaders of many emergency management agencies are appointees of the governor and, as such, may be in place for only a short period. These factors can be identified and interpreted through a study of the agency over time.

All affected groups, organizations, and agencies do their best to constantly improve their disaster response and recovery operations. Organizations usually have more difficulty responding to new requirements generated from outside their areas of control than if the change originated from within the organization. However, all agencies attempt to meet the legal, regulatory, and policy requirements to the best of their abilities.

Disaster response and recovery organizations are dedicated to helping people in time of need and do so to the best of their ability and in accordance with all relevant legal and organizational requirements. Changing the laws and requirements does not change the fundamental mission or character of the organization. Most individuals involved in disaster response and relief are involved because of their care and concern for people. They do the best they can with the resources they have to meet the needs of most of the people they serve.

With respect to the future of natural disaster relief, it is conceivable that some voluntary organizations may have to modify or eliminate their participation in natural disaster response or recovery activities if the legal requirements change to the point that they can no longer meet them, given their resources.

A theoretical example may illustrate this principle. Suppose it was required that an organization providing a shelter for humans had to provide a facility for pets at the same location. It is possible that a church that is presently willing to shelter humans would not want a kennel established in their buildings. At that point, they may decide to no longer shelter humans.

In addition, as legal and regulatory requirements grow, the number of facilities that might meet those requirements may decrease to the point that the available resources become insufficient to meet the need. Litigation and individual rights also have led to changes. We tend at times to forget that we are in these situations because of a disaster, and we must do the best we can for people in order to minimize suffering. It is not reasonable to be fixated on the problem that we do not have the recommended required square footage in a shelter. We must take care of the immediate needs to minimize the impact first and foremost on human lives and then deal with minimizing damage to property.

Laws Associated with Natural Disaster Relief

The most important law associated with disaster relief is the Robert T. Stafford Act. The act provides the authorities under which the Federal Emergency Management Agency provides major disaster and emergency support to both units of government and to individuals and families. The act is currently under review, and I believe it will see some significant changes. Most items included in this publication will be outdated in the very near future, so it is important for readers to reference these changes and understand that current law and regulations are always evolving.

The following is a list of laws that also involve disaster relief. Each is important in its own way.

Federal Laws

1. The Homeland Security Act of 2002, Public Law 107-296, 6 U.S.C. 101 et seq., November 25, 2003: This act established the U.S. Department of Homeland Security as an entity and incorporated the Federal Emergency Management Agency as part of the new department.

2. The Robert T. Stafford Disaster Relief and Emergency Assistance Act, as amended, 42 U.S.C. 5121-5206: This act established the means by which federal natural disaster assistance could be provided to the states and local jurisdictions to better provide aid to their citizens. When amended in 1988, the act established the Federal Emergency Management Agency. The act was further amended by the Disaster Mitigation Act of 2000 and the Pets Evacuation and Transportation Standards Act of 2006.

3. The Public Health Security and Bioterrorism Preparedness and Response Act of 2002, Public Law 107-188, 42 U.S.C. 247d: This act is primarily related to public health preparedness and response as they relate to acts of terrorism. However, the preparedness and response activities involved in this legislation serve only to improve public health response to other hazards, including those that occur naturally.

4. The Defense Production Act of 1950 (DPA), as amended by P.L. 102-558, 106 Stat. 4201, 50 U.S.C. App. 2062, and the Defense Production Reauthorization Act of 2003: This act was part of a larger civil defense and war mobilization effort in response to the cold war. This act provides the president with three powers: the power to require businesses to sign contracts or fulfill orders deemed necessary for national defense, the power to establish mechanisms to allocate materials and other resources to promote the national defense, and the power to control the civilian economy to ensure availability of critical materials for defense needs.

5. The Economy Act, 31 U.S.C. subsection 1535 & 1536: This act was a measure to balance the federal budget in 1933. Some of the effects of this act are seen in the requirements regarding individual and public assistance programs under the Robert T. Stafford Act.

6. The Posse Comitatus Act, 18 U.S. 1385: This act generally prohibits federal military personnel and units of the National Guard under federal authority from acting in law enforcement capacities within the United States.

7. The National Emergencies Act of 1976, 50 U.S.C. 1601 et seq.: This act provides several important "procedural formalities" that limit federal emergency response, including the time period a state of national emergency can last and executive powers relating to revocation of habeas corpus and the right to a grand jury for National Guard members when on federal status.

8. The Comprehensive Environmental Response, Compensation, and Liabilities Act (CERCLA) as amended by the Superfund Amendments and Reauthorization Act of 1986, 42 U.S.C. 9601 et seq., and the Federal Water Pollution Control Act (Clean Water Act), as amended, 33 U.S.C. 1251 et seq.: This act was created in an attempt to protect people, families, communities, and others from heavily contaminated toxic waste sites that have been abandoned. Though this act does not directly affect response to natural disasters, it does have secondary impacts.

9. The National Oil and Hazardous Substances Pollution Contingency Plan (NCP), 40 C.F.R. Part 300: This document describes the federal government's response to oil and hazardous materials spills.

10. The Cooperative Forestry Assistance Act of 1978, Public Law 95-313: This act provides for federal assistance and interstate assistance regarding forest fire emergencies.

11. The Communications Act of 1934, 47 U.S.C. 309 et seq.: This act overhauled the federal oversight of all forms of communications. As

such, it regulates how communications assets can be used during emergency situations.

12. The Insurrection Statutes, 10 U.S.C. 331-334: This act governs the president's authority to deploy federal military personnel within the United States to put down lawlessness, insurrection, and rebellion. The effect of this act on natural disaster response is in the area of crowd control relating to disaster operations, specifically to protests or other similar gatherings.

13. The Defense Against Weapons of Mass Destruction (WMD) Act, 50 U.S.C. 2301 et seq.: This act, though not directly related to natural disaster response, provides for federal assistance in response to nuclear, radiological, chemical, and biological weapons used against the United States. Many of the provisions within this document can serve as guides for other types of support relationships.

14. Emergencies Involving Chemical or Biological Weapons, 10 U.S.C. 382: This act provides for stockpiling resources and their use during response to weapons of mass destruction, specifically chemical or biological weapons. These same assets can be made available for other hazard responses, with certain conditions in place being met.

15. Emergencies Involving Nuclear Materials, 18 U.S.C. 831(e): This code citation provides for certain restrictions on transactions involving nuclear materials, which could result in federal law enforcement investigation and federal response to the effects of an incident.

16. The Veterans Affairs Emergency Preparedness Act of 2002, 38 U.S.C. 1785: This act provides the authority for Veterans Affairs facilities and other resources to respond to emergency situations in their vicinity.

17. The Atomic Energy Act of 1954, as amended, 42 U.S.C. 2011-2284, and the Energy Reorganization Act of 1974, U.S.C. 5841 et seq.: This act provides the fundamental U.S. law on both the civilian and the military uses of nuclear materials.

18. The Price-Anderson Amendments Act, Public Law 100-408: This act provides certain liability protections for non-military nuclear facilities.

19. Furnishing of Health-Care Services to Members of the Armed Forces During a War or National Emergency, 38 U.S.C. 8111A: This act provides for health care to be afforded to members of the armed forces during war or during national emergency situations.

20. The Resource Conservation and Recovery Act (RCRA) of 1976: This act's goals are to protect the public from harm caused by waste disposal; encourage reuse, reduction, and recycling; and clean up spilled or improperly stored wastes.

21. Rural Development Act of 1972, Public Law 92-419, as amended: This act provides for response and recovery activities in rural areas, aimed primarily at infrastructure recovery.

22. Pets Evacuation and Transportation Standards Act of 2006—amends the Stafford Act regarding pet support: This act provides standards of care and sheltering for household pets, service animals, and companion animals during all events.

Recent disasters have caused major changes and continue to cause major changes in federal laws. As we continue to see lessons learned, we will see laws, programs, and other aspects of response change. Examples include the Pets Evacuation and Transportation Standards Act of 2006 (PETS Act). This is a post-Katrina amendment to the Stafford Act that provides requirements for state and local jurisdictions to plan for and provide shelter for pets near human shelter operations.

Along with the PETS Act, there are more stringent requirements for the provision of appropriate shelter for special needs populations. One of the most significant results of this legislation is that there has been a great deal of discussion on the definition of which populations are properly termed "special needs." Initially, this population's definition was centered on medical needs, language barriers, and institutionalized populations. This is now taking on a wider meaning to include pet owners and the like.

In addition, the appropriations bill for DHS last year included language to require DHS to include the governors of West Virginia and Pennsylvania in reviewing plans for a regional evacuation of the National Capital Region (NCR).

These laws provide state and local government the guidance on how they respond to the needs of the citizens by applying federal programs. The implementations of these laws are crucial. As with any law, interpretations by the agencies and people that implement them are key. In many cases, congressional intent is not followed, which results in additional changes to programs.

The primary "law" that is critical to relief programs is the Stafford Act with all of its amendments. It is essential to understand the various program areas within this act in order to provide the best possible support to affected populations. However, no individual can be a specialist in all areas of this act. It is too broadly scoped to assume that one person could acquire all of this knowledge and understand these nuances.

With respect to how relief advocacy and legal representation are affected, again I think it is not the interpretation of the law itself, but rather the programs generated by the law that are crucial. Most of the staff at the federal and state levels try to find ways to render aid to the citizens of this country and not ways to deny assistance.

Critical Information Regarding Natural Disaster Relief

There are many key elements in disaster response we could discuss, and your opinion here will vary based on who the author is and where they are located. There are many good existing programs in place today with a very dedicated workforce implementing them. We cannot let events like Katrina drive us to make rash decisions and abandon programs that have been successful. Knee-jerk reactions to events often lead to the development of bad laws and programs.

One of the most important factors in relief efforts to minimize the property damage and human suffering is preparedness and mitigation programs. Citizens should be as prepared as possible to take care of there own needs in the early stages of a disaster to give government agencies time to respond. Local and state governments should have their operations and logistical programs in place. Risk analysis should be completed, and any gaps should be studied and taken care of. Any needed mitigation efforts should be completed to help deal with reoccurring or potential problems.

Those managing a disaster must always have the flexibility to handle the event and not be hamstrung by laws and regulations that they know they must meet to receive federal funding. I strongly believe in the common sense approach. Because of the event, your activities, organization, or priorities can change in seconds. It is often necessary to manage multiple situations at a time, and to be successful, you must be flexible in your actions, reactions, and solutions to problems and events.

This is important because as we look to the future, we must look at prevention as being a key to disaster relief. A good comparison is the medical insurance community. If an individual gets sick, their medical insurance carrier will pay thousands of dollars for medical bills of a sick person, but that same carrier will not spend any money toward prevention programs that may have minimized that illness. This is evolving, and we need to evolve the same way in disaster response. If we can minimize the impact on people and property before an event, we can minimize our losses during an event, and everyone wins.

In terms of the future of natural disaster relief, we should look to place more emphasis and funding on pre-disaster programs, so we may be able to spend less on the response and post-disaster programs. The key is common sense.

Cases, Statutes, and Regulations

The Web sites and documents included in this segment provide a basic overview of the cases, statutes, and regulations that affect disaster relief. There are, in fact, numerous cases and statues that involve disaster response; therefore, it is impossible to list them all here. The topics of those cases vary greatly, from dealing with who may be responsible for a disaster such as a fire or flood to issues involving accountability, as currently can be found in cases related to Katrina. Some of these Web sites will yield sample contracts and guidance. Other laws are listed in other portions of the chapter.

Laws, rules, Web sites, and documents that apply to disaster relief include:

Insurrection Act TITLE 10 > Subtitle A > PART I > CHAPTER 15

Emergency Management Assistance Compact (EMAC): www.emacweb .org/

Federal Executive Orders and Presidential Directives

1. Executive Order 12148, Federal Emergency Management, July 20, 1979, as amended: Combined previously stand-alone federal organizations and agencies into the Federal Emergency Management Agency.

2. Executive Order 12333, United States Intelligence Activities, December 4, 1981: Directs all federal agencies involved with intelligence functions to cooperate with the Central Intelligence Agency.

3. Executive Order 12382, President's National Security Telecommunications Advisory Committee, September 13, 1982: Established the President's National Security Telecommunications Advisory Committee.

4. Executive Order 12472, Assignment of National Security and Emergency Preparedness Telecommunications Responsibilities, April 3, 1984: Describes the National Communications System's responsibilities for national security and emergency preparedness.

5. Executive Order 12580, Superfund Implementation, January 23, 1987: Although this executive order does not have direct impacts on natural disasters, it is very important to emergency management in an all-hazards approach. This executive order establishes the National Contingency Plan and assigns various federal agencies responsibilities for hazardous materials responses.

6. Executive Order 12656, Assignment of Emergency Preparedness Responsibilities, November 18, 1988, as amended: Provided for and assigned emergency preparedness responsibilities to various federal agencies.

7. Executive Order 12742, National Security Industrial Responsiveness, January 8, 1991: Requires the private sector to provide priority service to the federal government during emergency situations, whether disaster or other conditions, as set out in the document.

8. Executive Order 13284, Amendment of Executive Orders and Other Actions, in Connection With the Establishment of the Department of Homeland Security, January 23, 2003: Established the Department of Homeland Security.

9. Executive Order 13286, Amendment of Executive Orders, and Other Actions, in Connection With the Transfer of Certain Functions to the Secretary of Homeland Security, March 5, 2003: Reorganized the Department of Homeland Security to include transfer of the Federal Emergency Management Agency (FEMA) and a subordinate structure.

10. Presidential Decision Directive 39: U.S. Policy on Counterterrorism, June 21, 1995: Established federal counterterrorism policy.

11. Presidential Decision Directive 62: Combating Terrorism, May 22, 1998: Revised counterterrorism policy and activities.

12. Homeland Security Presidential Directive–1: Organization and Operation of the Homeland Security Council: Established the Homeland Security Council and eleven subordinate committees.

13. Homeland Security Presidential Directive–2: Combating Terrorism Through Immigration Policies: Established immigration policies aimed at reducing the potential for terrorists gaining access to the United States.

14. Homeland Security Presidential Directive–3: Homeland Security Advisory System: Established the Homeland Security Advisory System,

a series of color-coded conditions and related activities targeting reduction of hazards from potential terrorist activities.

15. Homeland Security Presidential Directive–4: National Strategy to Combat Weapons of Mass Destruction: Established multiple methods for combating the proliferation of weapons of mass destruction with the aim of securing the United States from attacks using such weapons.

16. Homeland Security Presidential Directive–5: Management of Domestic Incidents: Established the National Incident Management System (NIMS).

17. Homeland Security Presidential Directive–6: Integration and Use of Screening Information: Established policies for the development and use of information regarding individuals to provide additional safeguards for the United States through a wide range of agencies and activities.

18. Homeland Security Presidential Directive–7: Critical Infrastructure Identification, Prioritization, and Protection: Established a national policy for federal departments and agencies to identify and prioritize U.S. critical infrastructure and key resources and to protect them from terrorist attacks.

19. Homeland Security Presidential Directive–8: National Preparedness: Established policies to strengthen the preparedness of the United States to prevent and respond to threatened or actual domestic terrorist attacks, major disasters, and other emergencies by requiring a national domestic all-hazards preparedness goal, establishing mechanisms for improved delivery of federal preparedness assistance to state and local governments, and outlining actions to strengthen preparedness capabilities of federal, state, and local entities.

20. Homeland Security Presidential Directive–9: Defense of United States Agriculture and Food: Established a national policy to defend the agriculture and food system against terrorist attacks, major disasters, and other emergencies.

21. Homeland Security Presidential Directive–10: Biodefense for the 21st Century: Established policies and activities to protect the United States from attacks involving biological weapons.

22. 44 CFR Part 206 contains regulations governing the federal disaster recovery programs administered by FEMA.

Governor Joe Manchin appointed Jimmy Gianato director of the West Virginia Division of Homeland Security and Emergency Management in September 2005. In his capacity as director, Mr. Gianato has operational and planning responsibility for the state's response to all emergency and disaster operations and consequence management for incidents involving weapons of mass destruction and terrorism. During federal disasters, he serves as the governor's state coordinating officer with the Federal Emergency Management Agency.

Before assuming his present duties, Mr. Gianato was involved in emergency response for more than thirty years. He has served as a 911 director, an emergency management director, a firefighter, an emergency medical technician, and a law-enforcement official. He has received training in incident command, the National Incident Management System. He is a certified fire, emergency medical service, and law enforcement instructor and has successfully managed numerous nationally declared disasters in McDowell County and has been deployed to other parts of the state to assist with disaster efforts during times of need.

Prior to accepting his current position, Mr. Gianato was director of emergency services and 911 for McDowell County for fourteen years. He was assigned to the West Virginia State Fire Marshal's Office as an assistant state fire marshal for nine years and has been involved in numerous state and federal investigations.

Mr. Gianato has been a presenter for several organizations and conferences, including the 2005 All Hazards Forum, NENA, the American Business Club, the International Coal Symposium, and the West Virginia Coal Association, to name a few. He intends to continue to attend training and learn of new and better ways to improve the Division of Homeland Security and Emergency Management in the state of West Virginia.

Dedication: *I wish to acknowledge the staff of the West Virginia Division of Homeland Security and Emergency Management, whose tireless efforts on behalf of the citizens of West Virginia have resulted in the state's ability to minimize human suffering during times of crisis and who have contributed significantly to this text.*

Navigating the Complexity of Disaster Management

Joseph C. Becker

Senior Vice President, Disaster Services

American Red Cross

Coordinating a Response—People and Structure

Four key issues have framed much of the disaster response landscape in the post-Katrina era. The growing size, scope, and expense of disasters, the lack of citizen preparedness, the challenge of increasing service to people with a wider range of needs, and the escalating expense of preparedness have changed the landscape of disaster management. These issues are presented in this chapter from the view of the American Red Cross, but are intended to speak of the response community from the non-profit viewpoint.

The basic construct of disaster management remains unchanged. Initial leadership in a disaster rests with the local emergency manager. This person has great authority in times of crisis and has command of a variety of community resources. The emergency manager is also charged with coordinating the work of government and non-governmental agencies in leading an effective disaster response. Government agencies, utilities, schools, hospitals, law enforcement, the Red Cross, and (increasingly) the business sector come together in the emergency manager's operations center (EOC) to make decisions, share resources, coordinate activities, and communicate with citizens. This is a fairly predictable and consistent construct from disaster to disaster and from community to community. The structure has become even more uniform in recent years though a standard system called the National Incidence Management System (NIMS).

If the local government is overwhelmed by the scope of a particular disaster, additional resources can be requested from nearby counties in mutual aid and from the state. The state then requests federal assistance from DHS/FEMA (Department of Homeland Security/Federal Emergency Management Agency), which is responsible for providing and coordinating federal resources in large-scale events.

The American Red Cross responds to an average of 200 events a day across the nation. The vast majority of these responses are small—typically residential fires, weather-related events, and transportation accidents. We feed, shelter, distribute supplies, provide health and mental health services, and help meet emergency needs in the very small disasters and in the very large. In bigger events, such as floods and tornados, other non-profits step forward to serve, as well—typically faith-based groups, such as the

Southern Baptist Convention, the Salvation Army, and Catholic Charities, with each taking specific roles. In very large events like hurricanes and wildfires, many community groups assist, and the faith community in particular becomes a key contributor to the response. In the federal disaster plan (the National Response Framework), the American Red Cross is charged with integrating the efforts of the national non-profits in the earliest days of a disaster.

In addition to the federal government, there are local, tribal, and state disaster plans. The close coordination required at all levels is a challenge in normal times—and much more so in a disaster. The timely understanding of needs and integration of resources from a variety of sources is needed for an effective response. No community, state, or organization is resourced for a large-scale event; in a catastrophic event, it takes everybody.

Key Challenges

The growing size and scope of recent disasters are clear. While the frequency of natural or manmade disasters appears to be growing, the increased development of high-risk areas is putting a greater number of people in harm's way. The most obvious example is the population growth along the Gulf and Atlantic coasts, exposing millions to the threat of hurricanes. But we have also experienced tremendous population growth in our earthquake- and wildfire-prone areas. These increases and shifts in population have a very clear implication: when disaster strikes, more people are affected by it.

At the same time that our population is shifting into more disaster-prone areas, public expectations of assistance are increasing. Historically, disasters have been considered community events. Neighbor would help neighbor, and the community would pull itself together to recover with government and non-profit assistance. A frequent reaction now is a sense of "This bad thing happened to us, so what are they going to do for us?" In most cases, "they" refers to the response community writ large: local, state, and federal government, non-profits, as well as friends and family. There is a growing expectation among disaster victims, elected officials, advocacy groups, and businesses that the country will mobilize even more quickly to respond to disasters to restore citizens to at least their pre-disaster condition.

While the response community continues to build capacity and respond more effectively, a key step is the creation of a culture of citizen preparedness. We need to work to instill a sense of responsibility for ourselves, our families, and our neighbors. Each of us should have a plan, should assemble a supplies kit, and should know how to stay informed in a disaster. Tools and checklists for these three steps are available from both the Red Cross and government organizations. It is clear that in the earliest hours and days of a crisis, local first responders (police, fire, medical) are quite overwhelmed, and neighbor has to be able to help neighbor. We can each take the simple steps to make sure we have the food, water, supplies, and information needed for the first days of a disaster.

And yet we as a country have not improved our individual preparedness significantly, even given the very large events of the past decade. Red Cross data shows fewer than 30 percent of families have taken one of the three steps, and fewer than 10 percent of families have taken all three steps. This data has been tracked since right after 9/11, and the bad news is that we are barely moving the needle.

The Red Cross and government have put many resources toward family preparedness. And, while progress is slow, we have learned of two leverage points to change behavior. We know that an employer has great sway over its employees' family preparedness, and that children can influence parent safety behavior. If the employer says, "This business needs to keep going after a disaster. I need to know you will be able to come to work, so please figure out now—ahead of time—what your family will do and what it needs in an emergency so I can count on you," the data shows that behavior changes. We also know that children affect parent behavior. If the child comes home from school with a homework assignment to document the family disaster plan, families and communities become better prepared. The Red Cross has focused a lot of attention on schools, helping children learn preparedness and families prepare. Not only are more families taking the necessary precautions, but also we're educating the next generation, too, and building a culture of preparedness.

A third issue involves those of us in the response sector historically taking too simplistic a view of those who need help. In the past, planners and responders have categorized the affected population as either "general population" or those having "special needs." An example has to do with the sheltering of people during a disaster. The "general population" was presumed to be able to care for themselves as long as basic food, shelter, and supplies were provided. Those in the "special needs" category were presumed to have health care needs beyond what could generally be accommodated in a conventional shelter with care referred to medical facilities.

Having only two broad classifications of shelter residents greatly oversimplifies a complex situation. The Red Cross—and the entire response community—are moving toward recognition that there is a large spectrum of needs in any large body of people, and these needs must be better anticipated. Those needs might include specialized medical care, but they are just as likely to include transportation challenges, language barriers, disabilities, elder care issues, or family pets that need accommodations. People with disabilities, for example, have been too quickly seen as having special needs that can't be met in general shelters, when often they can remain with just small accommodations. In fact, most people in a disaster have some type of need that makes them unique, and government and the Red Cross need to be better equipped to address those needs. A good result of this work has been the inclusion of many more advocacy and service groups in disaster planning and response, groups that can help identify and serve people with a wide variety of needs.

Working Together

Large-scale disaster relief is not just the domain of a small, specialized group of agencies. While the scope of the disaster changes, the need for coordination grows. On the very small disasters, such as a fire, it is typically the local fire department and the Red Cross that arrive at the scene. But as the disasters get bigger, the number of organizations that can add value expands. In very large emergencies, many organizations traditionally not thought of as disaster responders step forward and want to serve. They can provide needed services—particularly to hard-to-serve areas—and are a welcome part of the community's response, even if spontaneously created and not in the county's plan.

Disaster relief agencies need to be prepared to effectively incorporate such assistance and to coordinate with unanticipated sources of help, as well as known partners. Our mindset needs to be very broad about who can add value and who can be brought into the process. Ideally, relationships should be in place before a disaster strikes, but there will almost always be help from unexpected sources. One of the great frustrations during Katrina was that so many businesses and organizations wanted to help but didn't know how. The tendency is to look for a large-scale programmatic solution, but the best answer—and often the simplest answer—is usually local.

An area that has been getting increasing attention over the past few years is the role of business. Business has the majority of assets in the country. It is important on a local level to connect private resources with local Red Cross and local government, to bring business into community disaster planning and response in an intentional way. The key challenge is to match the needs of those affected by disaster with the offers of donated goods and services from businesses. Forming partnerships, aligning resources with projected needs, and establishing methods of collaborating in crisis situations are all best done in the calm of advance preparation, rather than "in the moment." All too frequently, the response to a request of business, faith groups, and civic organizations to pre-plan is, "Call me when something happens, and we'll see what we can do." In short, good plans are never based on vague promises of support; they depend on solid, advance commitments that an emergency manager can rely on.

Resources

The American Red Cross tends to raise a substantial portion of the funds donated for victims of disasters. The majority of these funds are spent in the early phase (the response phase) to deliver direct services in what is called "mass care"—food, shelter, supplies, and emergency needs. When the public sees the work that the sector is doing, particularly in a large-scale disaster, the response is often quite generous and is often prompted by a desire to see work done for the victims' care. But long before the services were delivered, there was a cost of readiness—a cost to have volunteers recruited and trained, and supplies, vehicles, and equipment ready to roll.

The costs to maintain warehouses, IT systems, supply chains, and call centers are not typically considered, yet are a cost of readiness. Larger threats and greater expectations translate to higher expenses to build and maintain the capacity to respond. An analogy is with the local volunteer fire department. If firefighters respond to a family's burning house or medical emergency, many would be very willing to pay for the cost of that service. Few would have an issue paying for the volunteer fire department to respond and put out a fire. Its response begs the question of how the fire truck, the firehouse, the supplies, the workers, and their training all were made possible. Commitment to an effective relief system is not only supporting the direct services, but also supporting the systems, training, and logistical resources that it takes to deliver a comprehensive response. As the size of disasters has increased, the cost of readiness has escalated, too.

Of course, direct services also cost money. In the earliest days of Katrina, while the complete scope was not yet known, it was clear that Red Cross early response costs would be well in excess of $1 billion—a daunting sum to a non-profit. Clients were in need, and the organization had no choice but to borrow the money to initiate the response—seeking a $1 billion line of credit to continue service. It was a huge gamble, and it placed the organization's viability at tremendous risk.

The American Red Cross is chartered by Congress to provide a nationwide system of disaster relief, but funded by private donations; as a result, we cannot pick and choose disaster operations based on our financial situation, our ability to deliver service, our available resources, or the willingness of volunteers to serve. When disaster strikes, the Red Cross responds first, and then trusts that people and businesses will join the effort. If the Red Cross must borrow the money to deliver service, we do. In effect, our leadership trusts that America will value the service and respond with generosity.

To date, hurricanes Rita, Katrina, and Wilma have cost the Red Cross about $2.2 billion—an amount almost twenty times larger than the largest prior natural disaster response. More than $400 million was borrowed before the generosity of Americans helped us repay the line of credit and cover the remaining costs. The cost of the service provided by Red Cross volunteers was donated. The challenge going forward is building and maintaining the capacity required for the next catastrophe. Going back to the fire department analogy, the response was financed, but it was clear that the capacity to respond to a catastrophic event must continue to grow and must continue to be financed.

Federal/State/Local Resources

The Robert T. Stafford Disaster Relief and Emergency Assistance Act (Public Law 93-288) is the primary federal legislation associated with disaster relief. The Stafford Act generally allows the federal and state governments to spend the funds they deem necessary to mount a response. Costs for response are generally shared with the affected states. Typically, funding is not the constraint in government response; instead, the issue is one of coordination.

Under the Stafford Act, the president's declaration of a disaster area triggers a flow of federal resources into a community. Subsequent to Katrina, additional legislation was passed to permit the federal government to move even before a state requests the resources. In addition, states no longer have to demonstrate that they are overwhelmed before the federal government sends relief.

On the state and local level, additional legislation regulates emergency powers. In an emergency, the local emergency manager can declare the need for public resources and take possession of these assets for alternative uses (e.g., declaring a school a shelter). There are also different authorities granted through state and local legislation to local officials in times of disaster.

Change

The lessons learned from the past few years are many, and change has been prompted across NGOs, the private sector, and government. The Red Cross, for example, learned much in the Katrina response. The organization quickly increased supplies and capacities prior to the 2006 hurricane season. We have reached out to a wide variety of partners, bringing them into national and community disaster planning and coordination responses. Specific efforts to reach out to partners to better serve the diverse community have worked well in subsequent relief efforts.

Changing government is a much more involved process, taking longer with a wider array of inputs and approvals. A good example is the release of the new National Response Framework in February 2008. The revision reflects changes mostly borne of Katrina and the storms of 2005, but it was simply not possible to release such a comprehensive document quickly under the public process that is required.

Conclusion

It is very clear that Hurricane Katrina, catastrophic as it was, will be dwarfed by future events. Consider just the hurricane scenario—quite a few more densely populated areas are at a similar risk. We will have events that will affect more people, cause more destruction, and cost more money in the years to come. America is better prepared to respond than it was ten years ago, or even two years ago, but it has a long way to go to be ready for what could happen.

Earthquakes, fires, floods, terrorist incidents, pandemics, and unforeseen incidents are in our future. The question isn't, "What will they do to get ready?" While "they" are building capacity, relationships, and readiness, "we" all have a role in this. It will take every American to prepare families, communities, and the nation—to build a culture of preparedness.

Joseph C. Becker is the senior vice president of disaster services for the American Red Cross, a human service organization in existence since 1881. The American Red Cross is dedicated to providing relief to victims of disasters and helping people prevent, prepare for, and respond to emergencies. Mr. Becker leads the organization's disaster relief. In this role, he led the Red Cross's two largest relief efforts to date—the 2004 hurricanes in Florida and the Hurricane Katrina response.

Mr. Becker joined the national headquarters staff on January 1, 2004, as the vice president of Response. Before assuming this role, he was the executive director of the Greater Carolinas Chapter of the American Red Cross, starting in February 1997. His Red Cross involvement started much earlier, as a member of the chapter board of directors from 1992 to 1996.

Prior to his employment with the Red Cross, Mr. Becker was part of the management group of Kings Entertainment Company. At the end of his twenty-three-year career with the company, he was the vice president of operations at Paramount's Carowinds.

Born and reared in Cincinnati, Ohio, Mr. Becker received a degree in business administration from Miami University in Oxford, Ohio, in 1979.

Preparation, Coordination, and Responsibility in Disaster Management

Barbara Farr

Director

Vermont Division of Emergency Management

The Current State of Disaster Relief

We live in a just-in-time society for food, banks, communication, media, and just about any service you can imagine. As a result, when a disaster occurs, there is an expectation that help is on the way immediately. While this expectation is met on most occasions under normal circumstances, it can be delayed when we experience a disaster like a flood or snowstorm or when first responders don't have the resources right away because of circumstances beyond their control, or because of the magnitude of the event.

That's why preparedness is considered paramount in disaster management; we just can't say it enough: What do you, as a business, as a family, or as an individual, need to do to be prepared? What sort of insurance policies should you have? Are they enough? Do you have any preparedness plans in place, especially if you are a person with special needs or on life support or disabled in any way? Do you know what you need to have and do to be ready to survive a disaster before help gets to you?

Many people with special needs believe that they will be taken care of automatically and immediately during a disaster. That is partially true, as first responders do focus on the most vulnerable, including elderly and those with special needs when organizing and carrying out evacuations. However, largely because of privacy laws, and the basic desire many have for protecting their privacy, we simply don't know where all of these people are.

Vermont Emergency Management found out how big this problem was just last year, when a downpour caused extensive flooding and power outages in Vermont. Several people with special needs were expecting help, but the local community was not made aware of their circumstances in advance; they weren't on any publically available database or list, so we had no plan in place to care for them. What's more, they didn't have their own plans in place. When we learned of these people and their needs, first responders immediately scrambled to make sure they were taken care of, but it caused questions to be asked about how we can do this better next time.

One specific example of this comes to mind where a single, elderly woman had Alzheimer's but wanted to live in her own home. Her daughter and son-in-law visited every day to make sure she was fed and had everything she needed. They locked her in at night because she was a wanderer. In April of last year, we had a large windstorm that hit the Alzheimer's patient's community particularly hard. It knocked out power, and it was immediately apparent that it would take a few days to restore service. The woman had to be moved, but the daughter and son-in-law didn't have room for her. The local shelters could not take her because they could not guarantee twenty-four-hour supervisory service, and they had a policy about not locking doors. The family didn't have a plan in place, and their options were limited when they reached out for help. The Alzheimer's woman ended up in a nursing home, where she received the proper care she needed, but two months later her insurance company disputed paying for the unscheduled nursing home visit.

This example shows that even in a short-term emergency, there are long-term repercussions. It all comes back to preparedness and expectations. The expectation is that everyone will be taken care of right away, but sometimes that is just impossible. The most important advice is this: be ready to take care of yourself and your family for a certain amount of time before responders can get to you. The recommended time in Vermont is to be prepared to be self-sufficient for up to seventy-two hours. Emergency responders do a great job and save lives every day, but during a large disaster, they just don't always have the resources to get to everyone right away. In Vermont, response in rural areas can be hampered because of road closures due to flooding, snow and ice conditions, trees blown down, and power outages.

Government Preparation

The disaster planning, response, and recovery network is extensive. It involves all levels of government, from the Federal Emergency Management Agency down to the volunteer fire department and emergency medical services in even the smallest communities. With such a wide array of skills and resources, it is important that everyone be on the same page, and that means relationships and training are vital in ensuring a seamless response. You need to know that even if the person you work with during the disaster is not the same person you work with on a regular basis, there is a universal structure in place that is proved to function as an effective response tool. That structure is the National Response Framework and Incident Command System (ICS).

In state emergency management offices, when we ramp up to respond to an emergency situation, we are not the actual on-the-ground responders. We are the coordinators of the overall response effort, and that is important to know when local and regional resources are exhausted. All disasters are local, so it's local first responders who are on the scene, and if the disaster exceeds their capacity, they bring in mutual aid, and when that is exhausted, the state emergency management office coordinates additional state and federal resources.

In emergencies, the non-profit sector, such as the Red Cross, United Ways, and church groups, is extremely important. In addition, there are volunteer organizations in each state called Community Emergency Response Teams (CERT) that were set up through Homeland Security and FEMA. The CERT volunteers are not first responders, but they do assist first responders. They might do traffic control while a scene is being taken care of, assist with shelter coordination, or clean up debris after a disaster, or they may even take care of pets that are left behind after a disaster. Through federal funds, our office provides them with the equipment and training they need to assist in times of disaster.

There are four main phases of any disaster: preparedness, response, recovery, and mitigation. The greatest portion of our time is spent on planning and preparedness. We have a state emergency operation plan and local emergency operation plans. We exercise those plans, and we train people for all types of events and scenarios. The extent of planning and exercising has increased tenfold since 9/11 and Hurricane Katrina. All of our efforts have increased our ability to respond and recover more rapidly and efficiently. We exercise and plan with our surrounding states and federal partners on a regular basis.

Vermont has a nuclear power plant within our borders, and the ten-mile planning zone extends into New Hampshire and Massachusetts. While these disaster plans are different from flood, hurricane, and terrorist disaster plans in that they are mandated by the federal government through the Nuclear Regulatory Commission, as a result those communities are much better prepared to address any hazard.

Recovery

When any disaster, like a flood or a hurricane, occurs, there is financial help available to communities and individuals, but it is not automatic. If public damage from a disaster reaches a certain financial threshold (in Vermont that threshold is approximately $1 million), it may qualify for a presidential declaration, and, if approved, the federal government will pay for disaster recovery to qualifying public property.

The declaration process starts with a gubernatorial-declared state of emergency, after which the governor applies for a federal declaration. There are two FEMA declaration levels for which a state may qualify—public and individual. Individual assistance reimburses private homeowners who have lost uninsured property. FEMA will reimburse them up to a pre-determined monetary amount. Public assistance helps make up for losses of public utilities or public facilities like roads, public buildings, and bridges. Debris management is another program for which a municipality can qualify. New Orleans after Katrina comes to mind as a situation where there was significant need for debris removal and significant federal reimbursement. Again, all of this aid is contingent on a federal disaster declaration; no declaration and you are on your own. The Small Business Administration (SBA) also has a loan program for disaster victims.

Recovery from any size event is difficult for everyone because it means that losses have been experienced. In my experience it is often the smaller events where public and private property has sustained damage that are the most difficult because there is not enough damage to qualify for a Presidential Disaster Declaration. In a small rural state, local budgets are tight and always scrutinized, so that any unexpected emergency situation always stretches a community's tax dollars. Even more difficult are private properties and businesses that have uninsured or underinsured losses. This happens frequently, and more often than not, individuals will claim that they didn't know it could happen to them. It all comes back to risk assessment and preparedness. When disasters happen, they cause an immediate reaction for most people and communities to prepare and mitigate against the next possible threat. Unfortunately, this is often short-lived and is forgotten over time, then lapsing into complacency, and the cycle begins again.

Legal Aspects of Disaster Relief

The Federal Stafford Act governs the policies that detail reimbursements to states for disasters when a presidential declaration is in effect. Each state has its own legislation, as well. In addition, New England states have a mutual aid compact with eastern Canadian provinces. The International Emergency Management Assistance Compact Memorandum of Understanding was ratified by Congress and signed by President Bush in December 2007. This MOU will allow for reciprocity of professional license certifications and reimbursement of Canadian responders who provide mutual aid in the United States, and vice versa, when requested through the Compact because state resources are exhausted.

Another piece of legislation is EMAC—the Emergency Management Assistance Compact, which is state-to-state aid. When Katrina occurred, responders from all fifty states converged on the Gulf Coast area, and were reimbursed by FEMA. Vermont participated by sending National Guard troops and local first responders in addition to Vermont Emergency Management staff. A requesting state coordinates through the National Emergency Managers Association to send requests for resources via e-mail to all fifty states, and whoever has that asset responds with their pledge of resources. Through the EMAC administrative structure, a determination is made on which of these resources is closest, and that resource is requested to go to help. Anyone who has a professional license generally needs to be licensed by a state to practice in that jurisdiction. However, under EMAC a licensee can transfer his or her license from one state to another under emergency circumstances.

Most legal questions in disaster relief come down to deciding who will pay for a response and who is liable for damage. There are standard-of-care issues, insurance issues, mortgage losses, injury resulting from response issues, land-use issues, and the list goes on. There were unsafe formaldehyde issues in the travel trailers that were provided for temporary housing by FEMA. Every disaster has its resulting legal issues that are present during the recovery stage.

An ongoing discussion and unresolved legal issue surrounds land use and the takings issue. Rebuilding in areas that experience repetitive loss from flooding, earthquakes, landslides, and fires, and coastal barrier zones that are often affected by hurricanes come into question. Who should continue to pay for losses that are in high-risk areas—insurance companies, communities, landowners, or taxpayers? It comes down to land use policy and development or redevelopment in high-risk areas and the legal issue of whether to compensate the landowner in a buyout to prevent future loss.

The general public, businesses, developers, and investors need to be well-informed on what risk assessments are in the areas that they represent. All states and most communities have plans in place that detail the hazard inventories and risk assessments for their respective areas. Everyone should have their data backed up off-site and have continuity of operations plans in place. The sooner a business can get operational after a disaster, the better they will be to help others recover from a disaster.

Another issue that is becoming more of a concern involves the fine print in insurance policies. Many people or businesses do not read the fine print, so they do not realize they might not be covered for a type of situation that may affect them. In most cases, insurance companies carry the standard fire insurance, but with any kind of water damage, they typically have exemptions, especially for flood damage. If properties do not have special flood insurance through the National Flood Insurance Program, they may not be covered for any flood damage. Whether it is a business or private property, most people believe a disaster will not happen to them, and when it does, they learn quickly that they will not be made whole by the federal government.

Changes and the Future

Natural disasters, particularly recurring problems like floods, coastal damage, hurricanes, and tornados, are bringing the issue of redevelopment in appropriate places to the forefront. We have seen that in New Orleans. Do they redevelop the Ninth Ward, or do they turn that into park space?

What do you do with coastal areas where the ocean is encroaching on development? The problem is compounded by development pressures to build or rebuild in inappropriate places. We have no control with Mother Nature and our changing environment.

Changes in attitudes and policies occur with each new devastating event. In recent history catastrophic events, such as the 9/11 World Trade Center terrorist attack, the Indonesian tsunami, Hurricanes Katrina and Rita, devastating tornados and floods, and now melting glaciers and rising waters, cause changes in policy and perspectives to take place. With these recent events, there has been a lot of retooling and rethinking in how the federal government is involved in the all-hazards response and who is to take the lead. Additional changes take place with turnover in elections and administration of the policies. The war in Iraq has compounded the stretching of the federal budget for national preparedness and readiness.

Another change that has occurred is federal policy on rescuing animals due to the outcry and lack of a clear animal response component in emergency plans. This is new within the last couple of years and very much a large component of emergency management.

We are now seeing caps on eligible expenses in recovery, like debris management and hazard mitigation. The caps are a way of limiting expenses. There are different types of grants—individual assistance, public assistance, and Small Business Administration loans for businesses, homeowners, renters, and non-profits. If you do not get an individual assistance declaration, which is very difficult to get, most often the SBA will come in with its disaster loan program and loan money at low interest rates to recover. One area of discussion regards the fact that states do not know when or if they will be eligible for individual assistance. There are limited guidelines as to how many private properties need to be damaged in each disaster, and to what degree, to qualify for assistance, but it also can be very ambiguous.

Because of the work that my department does, we are careful to review federal policies that are proposed and determine whether there will be an adverse impact on the state. States are members of the National Emergency Management Association, and NEMA solicits from the states any impacts the proposed policy changes would have and acts as a collective voice for the states. We do have a very good network and committee structure set up with all the states in the entire country through this organization. Each year there is a federal funding proposal for emergency management organizations that is never enough to meet the demand of the public's expectation for preparedness, response, recovery, and mitigation. Through NEMA, states have a collective voice to reach our congressional delegates to stress how important this funding is. Through this network, states have been able to sustain and increase federal support, but it has never been to the level where it adequately addresses the need.

There are many different agencies involved in disaster response and relief, and Congress is the main driver in guiding policy and how these different agencies respond. Within the National Response Framework structure, there are fifteen Emergency Support Functions (ESF). It is a plan that goes from the federal government to the state to the local community, or the local community to the state to the federal government, so it can go up or down the government chain, but everyone is under the same structure. The first ESF is Transportation. Transportation in almost any emergency is a critical component. In Vermont, we utilize our Agency of Transportation to help with road closures, flooding, failed culverts, etc. The other fourteen ESF divisions include everyone from Agriculture to Law Enforcement to Fire Safety to Public Information. Agriculture and the Department of Health are major partners in emergency response and coordination for us; Public Service, which oversees utilities, is another one. When power outages occur, which they do all the time in a natural disaster, and even in manmade disasters, Public Service is another important support function in making sure the critical infrastructure concerning utilities is up and running as soon as possible.

There are political factors involved in natural disaster relief at the local, state, and federal levels, and they all interplay during any kind of event. If something goes wrong, no matter what, there is often some finger-pointing. We certainly saw that after Katrina. It can happen at the federal level, the state level, and the local level. I believe each state has a very high level of understanding of what needs to be done in emergency situations and how it needs to be done, but quite often, when different agencies of the federal government come in, there can be political maneuvering that complicates the response and recovery effort. For this reason, there has been dramatic improvement in the multi-partner approach to emergency management, coupled with an increase in exercises and training across disciplines that has brought many new alliances to work together.

Two organizations really are watching what is going on with policy changes in natural disasters. One is NEMA, the National Emergency Management Association, and the other one is IAEM, International Association of Emergency Managers. They are both excellent organizations. The first is primarily geared toward state directors, and the second is geared toward all emergency managers at the state, local, and county levels. These two organizations are important in obtaining information that is accurate, consolidated, and timely. They are accountable and excellent advocates for emergency managers in the field.

Improvements are being made every day in the emergency management field. Awareness levels through real events and the media have been helpful in highlighting the need for preparedness and accountability of one's own actions. Enough cannot be said about the benefits of preparedness and mitigation prior to a disaster event. The better we as a society address our risks upfront, the less of an impact we will experience when disaster does strike.

The transition to all-hazards since 9/11, and now with the potential for a pandemic, brings the different response partners together to provide a seamless, coordinated effort for response and recovery. We need to adapt to meet the challenges of our changing environment, whether the changes are rising oceans, overpopulation, terrorist threats, reliance on foreign countries for food, oil, and other products, overdevelopment, disease, or political unrest. How well we prepare and adapt will dictate how well we survive the small and large disasters that occur on a regular basis. I feel that we have made great strides in preparing the general population in Vermont through outreach and education, but feel that this is an ongoing challenge. When emergency situations do not occur regularly, society becomes complacent and less cognizant of the potential impacts.

Barbara Farr was appointed director of the Vermont Division of Emergency Management in June 2005. She has been directly involved in emergency management issues for twelve years. Her oversight includes planning, preparedness, response, mitigation, and recovery for all-hazards for the state of Vermont. Developing and enhancing partnerships among the local communities, state departments, and federal entities in the emergency management field are a critical focus. She is the Region 1 vice president for the board of directors of the National Emergency Management Association.

Ms. Farr's previous experience includes ownership of a consulting firm specializing in emergency preparedness, recovery, and mitigation through FEMA and Homeland Security contracts and a former position as director of a Vermont regional planning commission. Ms. Farr led the county to recover from several federally declared flood disasters in the 1990s. She also was instrumental in securing millions of dollars in recovery funds through federal grant programs and through Project Impact, a former FEMA mitigation initiative. She has also worked for the National Institutes of Environmental Health Sciences in North Carolina and conducted research at the Woods Hole Oceanographic Institute in Massachusetts.

Ms. Farr holds a master's degree in environmental law from the Vermont Law School and a bachelor's degree from the University of New Hampshire.

Dedication: *I would like to thank Mark Bosma for editing and wordsmithing this chapter.*

Interdependency of Natural Disaster Relief

Arthur W. Cleaves

Emergency Management

The Current State of Disaster Relief

What makes New England unique in the area of disaster relief is that New England geographically is not as large as other regions in the country, but when a natural disaster strikes New England, it is likely to affect all six states. The cross-coordination of activities among the states is probably more critical, therefore, so there is a unique aspect to that. Geographically, we can respond to each of these states in a matter of hours because of their size. The bad thing is that there *will* be more than one state affected by a natural disaster, like a hurricane coming up and hitting Rhode Island, Connecticut, and Massachusetts, and then sweeping through all of New England. Responding to all the six states at the same time, and all the government agencies involved in that relief would make things complex.

The Patriots Day Flooding that happened on April 16, 2007, is a prime example. There was major flooding in all six New England states during that event; the event was considered a major Nor'easter that happened on the day of the Boston Marathon. There was enough damage in each of the six New England states that the governors requested major declarations, and President Bush granted those requests for public assistance for all six states and individual assistance for three. There was significant damage to homes and properties in Connecticut, New Hampshire, and Maine. In New Hampshire, more than 1,000 individual homes were affected by major flooding.

I think the most overarching concern in disaster relief right now was a result of Hurricane Katrina. The Katrina Reform Act that Congress put in place really looked at rapid and compassionate relief for a disaster victim. That is the largest concern—how to get into the scene of a disaster and give relief and how to make it timely. I think much of it was born out of the graphic pictures we saw following Katrina. Probably the one largest concern is how you bring that relief the most cost-effective way. I would underscore the need for understanding the full spectrum of an emergency management system, from the local community, to a county and then regional level, and finally to a state level, and then the federal government when they have to come in with relief, and the responsibilities within all that chain.

All emergencies happen initially at a local level—the individual or the families or the individual community. Your first responders are instantly engaged in an activity that is not something they practice every day. This is quite different from a fire or an ambulance going to a scene. When it comes to emergency relief or natural disaster relief, you are responding to circumstances that are quite different. The workers are not necessarily trained; certainly, the skills are not honed; and not everybody understands their roles and responsibilities.

I think the greatest concern nationwide is individual preparedness. It is your responsibility to be prepared for seventy-two hours in case relief in a major event cannot reach you, if even local responders cannot reach you. This is something most people do not often think about. I am a retired Army colonel. In 1999, I took on the state emergency management role for the governor for the state of Maine, and I started learning local and state government, the emergency management aspects, the responsibilities of each of the individuals within that, and what responsibilities lie with public officials. So much of what I would like to emphasize is the responsibilities of individuals, and then public officials at all different levels.

I came up to the New England region two years ago, and now I am able to look at all six New England states and study the emergency management complexities and the differences in the structures. One state is not organized the same as another state. Then you go to the state level, and that is the first time you truly see a full-time organization of people who look at and plan for emergencies, usually under the governor's hand. Some of the large cities like Boston have an emergency management director who is full-time, but most communities do not have that position. The fire chief or one of the other officials within the community must plan for these events that happen very infrequently. I think often attention is not given to the disaster or the planning for the disaster because "It might not happen on my watch." The responsibility for public officials at the community level, and then the state level, is generally small, but they have to reach across all disciplines, and the connection is made back to the federal government.

Congress created a federal emergency management system that is the one discipline that reaches from the local community and creates a chain of command all the way up to the governor, and finally to the federal level, so you can reach all the way to the president during times of emergency. I do not think there is anywhere else in government where there is a group that cuts across all those levels.

Funding is driven by grants. We have an emergency management performance grant that is given each of the states, and they have to match half of it for salaries for employees who do the planning for natural disasters and/or manmade disasters. Many changes came after 9/11, looking at the terrorism aspect. There is not much difference in the results of a terrorist activity or a natural disaster. You have a consequence, and then you have to have the same type of relief for the individuals stricken or hit by whatever the disaster is.

Allocation of resources is very problematic. Each year, Congress allocates so many billions of dollars for disaster relief. How do you manage it so it meets the expectation of the individual who is stricken or the community that needs to rebuild infrastructure? There is a good chain able to quickly get the funding down to those levels, but you saw the expectations of hundreds of thousands of people, all seeking that individual assistance in a timely fashion. How do you create that system that allows you to handle great numbers of people who need subsistence? How do balance that against what those individuals' personal thresholds are? And how do you manage getting the money out? That is a very complex network, and an area that FEMA (the Federal Emergency Management Agency) and we have put a great deal of effort into. It is very complex when you start from the congressional flow of money down for disaster relief. And then it is also an accounting system. How do you get it out and move it quickly to the local communities down to the individuals, and use that network that is put in place so you can handle the huge number of victims at the same time?

Again, going back to the Patriots Day Flood, 1,000 homes were affected, and relief came quickly in terms of financial support through the emergency management organization because there is a tried and proven system in place. As an example, mobile vans are dispersed through the local emergency management agency in these types of events. They are set up just like a local office to do business for those communities that don't have the office spaces to do the business. The individuals can go there, and staff is there to register them, and they receive funding checks right there. These work very well for the towns.

Education is needed for the public. There is a lack of understanding of the emergency management system in place. Most people just a few years back did not know what emergency management was. We need to educate citizens on what to expect and how to expect it, and educate public officials on what exactly is in that chain and what they can expect or not expect.

Since Katrina, FEMA is really a changed organization. It is being termed the new FEMA. We are trying to be that much more responsive, watching for disasters, anticipating and putting the chain of resources in place so we can get relief in as quickly as we can. If you understand the whole system, where the resources come from, and what the timing is on that, then both individuals and communities can be better prepared. We are prepared and educated in many areas of our own lives; for example, we know when we go to an airline what the expectation is regarding security. Not understanding the system, I think, is the greatest shortfall.

Considerations

The Stafford Act, created by Congress, provides for disaster relief and gives the terms and conditions where the federal government would get involved if a local disaster occurs in a community and overwhelms the community's ability to respond and/or pay for the response. Responsibility goes to MEMA (the Maine Emergency Management Agency), and they look at those circumstances and try to help from a host of mutual support from other surrounding communities. When resources at the state level are gone, and the governor cannot garner enough resources for the recovery or response phase, then it rolls up to the federal level, and that is where the Stafford Act comes in. If we reach certain criteria and the president deems it is overwhelming for both the local and state levels, and the state resources are not great enough, then it is declared a major disaster, and the Stafford Act comes into play.

The Stafford Act is now very much in the eyes of Congress, which is examining its effectiveness. Many changes have happened since the post-Katrina Reform Act, headed up by Senators Collins and Lieberman, who sit on the governmental affairs committee that oversees the Department of Homeland Security and FEMA. It was a thorough investigation to look at the changing of the laws. Many of the things that we put in place here in New England are born out of the Katrina Reform Act. Individual disaster assistance is just enough to maybe get the person or families back on their feet, not restore them to whole as they were before, and then insurance comes in. The public infrastructure is also in line with that law: Do the roads or bridges need to be rebuilt? The act identifies how the parts of the infrastructure that need replacement will be covered under the act.

Legal assistance would be required throughout the disaster relief process. There is not one situation that a lawyer would not be involved in. If you look at Katrina, as the housing was put in place and trailers were put in place for the victims, there was the formaldehyde issue that is still going on, where victims who were given trailers claim illnesses caused by formaldehyde in the trailers. There are legal issues at every turn of every disaster. There are legal issues that arise, whether it is just simple disaster relief for flood victims, and then any extenuating circumstances always have legal actions involved.

Lawyers who become involved in natural disaster relief need to understand the emergency management system, and need to know what to expect and what the capabilities are. I think also it is important to know timing on relief, and to think about the things that will tide you over for that period of time. Becoming involved in requiring public officials to have a system in place is important, along with staying on top of education.

The United States as a whole is quite disaster prone. Nationwide, there is always a disaster going on—fires, tornados, etc. New England receives every type of natural disaster that there is and anything can occur here, but we usually get it to a minor degree. So I think more than anything, what we need for the future is to strengthen the system of disaster relief. From the president to Congress, everyone is very concerned about getting rapid disaster relief to those victims. How to do that is the tricky part. What do we put on the ground? What is in place—power restoration, water supplies? How do we get those greater resources in? So it means understanding the types of hazards in your area that you might face, and making public officials very aware, even about the rare ones.

A lot of study goes into understanding the relief that will be needed, and then putting together that chain of supplies and the resource management that goes with it. That involves everything in the private sector, and their involvement so that they know the relief resources available, and understand what the gap is and what that need is going to be—if it is power supplies we need, additional people to work at a shelter, or more shelter capacity. Then we have to figure out how to stage those resources, where they arrive, whether there will be people coming in, like in Katrina. If you think about it, a second disaster is all of these people arriving, looking for places to be housed themselves while they bring relief to the victims that are already in a devastated environment with no place to go. So you even have to think ahead to where the workers will be housed. Disaster relief means putting that all together, understanding the gaps, then identifying needs and resources, and making sure the resources are in that chain.

Improving disaster relief to be rapid and compassionate starts with a basic understanding of the roles and responsibilities of all organizations, federal, state, and local, so that they know and anticipate a response. So you will know what I am going to do and what you are going to do, and you will know what a very deliberate action will be. Knowing the other players will tell you a great deal about how you can depend on them. Crisp communication between the organizations is vital. We have put in place a situational awareness. When I came to New England and was a state director, I was wired to everything that was going on. I received information on Department of Transportation road closures; I watched the river flows. Flooding is the most prominent natural disaster we have in New England, so I watched the river flows on a day-to-day basis, if they got anywhere near flood stage.

When I arrived in New England, there was no system for FEMA to keep an eye on the ball and the circumstances that could give you a signal. You can see a hurricane forming, so you keep your eyes on it, but what is going on in New England that we can keep our eyes on? We are watching river flows and weather conditions now. There was not a system in place prior to some of the Katrina Reform Act. Congress asked us to do things specifically different within FEMA nationwide, and they added some experienced people to the regions, so we knew what to look for and what systems to put in place.

The cross level of information did not exist before. If you were in any one of the six New England states, I had a good perspective of what was going on in my states, and a little about the bordering states to us in Maine, but very much in touch with any of the activity that is going on that could lead me to an event, even a Hazmat spill and similar circumstances. You watch so you can anticipate if both the federal government and the state emergency managers will have to swing into action to help give relief to local communities. So we are watching a whole host of activities now, and cross leveling the information so that it communicates back and forth between them.

Most troubling for us is that there are not full-time people in all these positions, so you are trying to get this information flowing up and down. Some of this is work that is evolving today, so there is better understanding of the cross-coordination that has to take place and the situational awareness today, as we continue to create the network. We are well served today by information management in terms of Web-driven systems. The information that you can get via e-mails and Internet traffic creates a network that allows us to access this information quickly, and you can keep your eye on the ball. The information gets cross-leveled via the six states, and it has river flow information on an automatic touch of a map. It will also show the weather conditions. The power grid, where we are always watching for those power outages and how the grid is flowing today, is managed by the electrical industry—but is that flow good and solid and safe? So we are watching many things in many areas.

We even monitor fuel supplies in the winter months, so we watch for an interruption in any of the flows. Last year there was a potential shortage in propane because of a rail strike. What that would mean is that homes would be potentially without heat. It ended very quickly, but that is an example of being able to look forward and see those conditions that could lead to response, and then more important, what response we would have to put in place. If it is restoration of power, that would mean putting in generators for a temporary power situation. If people are out of their homes for an indefinite period, do we have adequate shelters? How many of them are in a given area? Where is your shelter, and who is your fire chief, so you can be prepared? In the winter months, very cold and very concerning times, if you are out of power or fuel, it could be a disaster built on price alone, let alone a true shortage. Remember fuel rationing a number of years back? There are circumstances that would put us in those kinds of situations. Today the fuel supplies are what the industry calls just-in-time supplies, so depending on a boat coming in on time, they do not store much fuel. It is all just in time, so the dollars are not tied up in tanks and supply on hand.

We need to evaluate and make an assessment of a total emergency management system by each state and at a regional level. I look at all of New England. I have to intricately know the capabilities of the organizations of emergency management all the way down to and including the first responders, their capabilities, and their breakpoints. An assessment has to be done first of all to see where they stand today, and then very much a systematic placing of the gaps if I know what the shortfalls will be. If a community does not have certain capabilities, when will they be overcome, and when will the state not have relief to bring in? We measure those gaps, so that is an assessment phase that is constantly ongoing and redoing every year as the emergency management performance grants or the Homeland Security Grants are given to the states, so that officials use that in the best capacity they can to build that response, whether it is a Homeland Security event or a natural disaster.

We look at the situation with an all-hazards approach, so how do you put in place that system, manage it, and then manage on a regional basis so that one state can cover for the other? What we hope to implement here in New England is looking ahead three years at what we can put in place. What I have described is really about partnerships, true partnerships that go beyond just an exchange of a business card, to understanding each other's roles and responsibilities. I term it as a seamless relationship between levels of government. If you put in place a seamless system, that means we would get over the barriers of the federal government being the bad guys and the states being independent government entities.

We have a common interest in disaster relief, so there is no turf involved. It requires understanding the roles and responsibilities that each level brings. Then we need to teach a culture of preparedness throughout New England at all levels of government, at an individual level as well as every community, and give it the attention that it should have, balanced with all the other issues. Then there need to be interdependencies between organizations—whether it is EPA or the Coast Guard or FEMA, we each have a vested interest, and we need to know we are interdependent on each other. Preparedness means depending on all of those major organizations being intertwined and interdependent on each other.

Arthur W. Cleaves was appointed regional administrator of the U.S. Department of Homeland Security's Federal Emergency Management Agency (FEMA) Region I Office in Boston, Massachusetts, in May 2006. In this capacity, he is responsible for coordinating FEMA emergency preparedness, mitigation, and disaster response and recovery activities in the six New England states: Connecticut, Maine, Massachusetts, New Hampshire, Rhode Island, and Vermont.

Prior to joining FEMA, Mr. Cleaves served as the director of the Maine Emergency Management Agency (MEMA) beginning in 1999. In addition to emergency management, Mr. Cleaves took on the role of director of Homeland Security for Maine during his tenure at MEMA.

Mr. Cleaves spent more than thirty years in the Army National Guard, rising to the rank of Colonel before retiring in 1999. During his time in the Army National Guard, Mr. Cleaves received numerous awards, including the Army Commendation Medal, the Meritorious Service Medal, and the Legion of Merit. Prior to his military service, he was in private industry, creating and heading a very successful auto parts business in Northern Maine.

Mr. Cleaves received a bachelor's degree from the University of Maine. During his military career, he completed the Command and General Staff College in Kansas, and the Joint Services Military Planning College at the Naval War College in Virginia.

An Ohioan's Perspective on Natural Disaster Relief

Nancy J. Dragani
Executive Director
Ohio Emergency Management Agency

Current Concerns in Natural Disaster Relief

Declaration Process and Small States, Rural Areas Issues

There are several issues that emergency management directors and FEMA (the Federal Emergency Management Agency) are struggling with across the nation. One topic that draws out strong opinions from coast to coast is the eligibility criteria for disaster declarations, particularly for rural states, as well as rural areas within urban states.

Ohio is a large state, seventh in the nation in terms of population. However, parts of Ohio are very rural. For Ohio to meet the state per capita requirement to be eligible for a public assistance disaster declaration, we need more than $13 million in eligible losses. We could have six or seven rural counties that have significant public assistance losses and meet or exceed their county per capita requirement, but because we can't meet the threshold required of the state, they might not be eligible for federal public assistance.

In addition to criteria for small states and rural areas, there is considerable discussion over stated criteria as found in 44 CFR (Code of Federal Regulations) versus unstated, unwritten criteria that emergency management directors and staff believe is often used by our federal partners. It is the unstated criteria that are subject to speculation and distrust. This perceived lack of transparency on the declaration process leads many state officials to question the impartiality of the process.

Transparency in process goes back to the declaration process, to a large degree. Right now, for some very good political reasons, there is not a lot of transparency in the process. Once the declaration request that goes from the state to the FEMA region leaves the FEMA region with a recommendation and goes on to headquarters, there is not a lot of feedback. It is as if the curtain closes. To a large degree that is probably a necessary part of the political nature of declarations and the need for the ability of leadership at whatever level to grant the declaration or decline the request.

The concern of many directors is timely notification of the declaration, either concurrence or denial, with adequately explained rationale. This is particularly an issue for declaration denials. Typically, an approved declaration request is received in days, while a denial can take weeks or even months. Often, the denial comes with a simple explanation that refers to criteria in 44 CFR 206.35 (b) (1) that stipulates that the event must be so extreme that it has exceeded the ability of state and affected jurisdictions to respond. Arguably, if a state requested a federal declaration, then the state and the affected jurisdiction did in fact believe that it exceeded their ability to respond. A more detailed explanation of the reason for denial will not only assist government leaders in explaining to constituents why the assistance is not available, but it may also help them better determine appropriate requests for the future.

Public assistance thresholds are fairly well defined in 44 CFR, 48 (a) (1), which indicates that FEMA will look at several factors to evaluate the need for assistance under the public assistance program. FEMA uses a figure of $1 per capita as an indicator that the disaster is large enough to qualify for federal assistance. In addition, FEMA has established a minimum threshold of $1 million in public damages per disaster. In addition to the state per capita, they evaluate the impact of the event at the local level because there may be disasters that cause a significant impact on local government, yet not meet the statewide per capita, that warrant federal assistance.

Individual assistance disaster declarations appear to be much more subjective. They can be based on impact, numbers of destroyed and severely damaged homes, or media interest. Again, small states and rural areas present difficult challenges for emergency managers at all levels. In many of Ohio's smaller counties, even meeting the Small Business Administration declaration threshold of twenty-five homes or businesses with more than 40 percent uninsured loss is difficult. Yet, to a rural county with 13,000 residents, a flood that destroys thirty homes and leaves another thirty with major damage is a *major* disaster. It makes a huge impact on that county, yet may not receive a major disaster declaration. In Ohio, we could have an event that is devastating to several small counties and not come close to meeting the threshold, so this is not just a small-state issue, but a rural area within an urban state issue, as well.

While 44 CFR 206.48 (b) (1-6) acknowledges that there is no set threshold for the Individual Assistance program, it refers states to a chart that indicates the average number of homes that have sustained major damage or were destroyed between July 1994 and July 1999. Based on the size of the state, the numbers range from a low of 173 in the small states (population under 2 million) to a high of 801 in the large states (population more than 10 million). In addition to the chart, 44 CFR refers to other factors, including concentration of damages, large numbers of injuries and deaths, large-scale disruption of normal community functions, and wide-scale loss of power or water, and impacts on special populations and voluntary agency assistance. There are many cases where other impacts besides sheer numbers of damaged properties seem to warrant a major disaster declaration, but the perception is, the numbers (of damaged or destroyed homes) must be apparent before a declaration is issued.

Even the most populated states could have an event that is devastating to several small counties and not come close to meeting the threshold, so this is not just a small-state issue, but a rural area within an urban state issue, as well.

I believe FEMA recognizes this issue and is attempting to address it in a variety of ways. It recently hired Brock Bierman, as their small state and rural advocate. A key part of his stated mission is to ensure that the needs of rural communities are addressed in the disaster declaration process and to assist small-population states in preparing their requests for disaster declarations. While this effort is still in its infancy, I think there is clear recognition that this issue needs attention. Certainly it is on the radar of both the states and FEMA. The question is how we can collaboratively work together to create a more transparent process, while recognizing that the federal government will always maintain some ability to be discretionary. We must also ensure that as we refine these processes, we don't let our most vulnerable constituents fall through the gaps. A major flood in rural Alaska or west Texas that destroys five or ten homes is no less devastating to those residents that a major flood in Cleveland, Ohio, that destroys 500 homes.

Many states have begun to address small disasters by developing their own public and individual assistance programs. All have unique characteristics and criteria. In some states, these are programs to provide immediate assistance while waiting for the federal declaration. Ohio has both an individual assistance program and a public assistance program that are for disasters that do not meet the FEMA threshold. They are activated only if we are confident that the damage will not meet FEMA criteria.

Ohio's public assistance program mirrors FEMA's public assistance program and generally allows the same type of reimbursement. To be eligible for reimbursement, applicants must have eligible disaster damages that exceed one-half of 1 percent of the applying jurisdiction's operating budget. The state will then reimburse 75 percent of the eligible costs.

The application packet can be found at www.ema.ohio.gov/PDFs/ApplicationPacket.pdf.

Ohio's individual assistance program is patterned after a combination of FEMA's housing and other needs assistance programs. The state must first receive an SBA (Small Business Administration) agency only declaration, which means that at least twenty-five homes or businesses have more than 40 percent uninsured loss. Once that occurs, the governor can activate the state individual assistance program for constituents who have unmet needs after they have applied for insurance reimbursements and/or an SBA loan.

For example, a flood occurs in rural Harden County, Ohio. It doesn't meet federal criteria for a major disaster declaration, but there are clearly local governments and constituents who need assistance. The state individual assistance and public assistance programs allow us to provide grants to individuals who are affected, as well as grants to governments that have incurred costs or damages because of a disaster, once our internal thresholds for each program have been met.

There are valid concerns from some states that strong state programs may result in a decrease in federal assistance under the assumption that state's don't need federal aid. The reality is most states are attempting to either assist their affected population with immediate needs while waiting for a declaration or, in Ohio's case, trying to provide for those governments and constituents who may have needs following small events.

State and Federal Partnerships

In Ohio, we have a solid working relationship with our FEMA region and many senior leaders and staff at FEMA headquarters. I receive feedback from region staff and have no qualms about picking up the phone and asking for details or insight on disaster denials. To the extent allowed, they are responsive and willing to offer advice and guidance. I feel strongly that my state organization should partner will FEMA as much as possible, recognizing that we serve different leaders and may not agree on every issue and position.

However, mutual respect and consideration and a clear recognition that ultimately we are serving the same constituents go a long way toward an effective and rewarding partnership. Unfortunately, I realize that this posture of partnership and shared commitment is not as strong in other areas of the country. Because of this, transparency in the process is more of a national issue than one we wrestle with in Ohio.

Catastrophic Planning: What Makes Sense?

Catastrophic planning is a hot topic, primarily because of a one-size-fits-all approach. Post-Katrina, states were asked to, in effect, plan for a "Katrina-like" event, particularly focusing on the evacuation and sheltering of the entire state. In Ohio, there are few, if any, scenarios that would drive us to evacuate the entire state, or even half of the state. It is a hard stretch from a risk analysis perspective to come up with what that scenario looks like.

Most EMA directors agree that we need to be prepared for major events that go beyond the routine disasters we face nearly every year, such as tornado, flood, and winter storms. However, Ohio doesn't need to prepare for all of the same events that Louisiana needs to be prepared for (hurricanes, for example) nor does Louisiana need to prepare for blizzards. This planning emphasis on catastrophic events has caused widespread concern that we are planning for a Katrina-like event in areas of the nation that will never face a Katrina, at the risk of not planning for those events that we know will occur.

If not this year, then next, Ohio will have flooding on the Ohio River. If not this year, sometime in the next five years, Ohio will have a major tornado and severe winter storms. We must continue to plan for and improve our responses to those types of known threats, while continuing to take into account a catastrophic event. We should not make the mistake however, that by planning for the catastrophic event will ensure that I can respond effectively to the routine. I think that the argument that if you are prepared for the worst case, then you are prepared for the garden-variety event does not hold water. Since 9/11 and Katrina, we have been planning, training, and equipping for terrorist events and catastrophic events (read hurricanes) respectively, often at the expense of our known threats and hazards.

Disaster of the Hour, or "What Are We Planning for Today?"

"Flavor of the day" is a huge issue for us, and by that, I mean unexpected or large-scale disasters, such as the Virginia Tech shooting or Hurricane Katrina or 9/11. Obviously, we must learn from events and be flexible enough to shift when appropriate, but this must be balanced with ongoing work efforts. September 11 obviously shifted our entire focus away from natural disasters to terrorism, at least until Katrina. Then our focus shifted to catastrophic planning and the issues that were particularly problematic during Katrina, such as mass evacuation and mass sheltering. Virginia Tech drove a lot of activity on whether colleges and universities are prepared for an active shooter and processes to notify students and faculty of an event on campus.

It often seems that every significant or unusual disaster, particularly one with significant media coverage, causes a change in focus. Consequently, the people at the ground level trying to get the job done get about 70 percent done with one focus area, and then are forced to shift focus and work on another issue. This creates frustration for folks who are trying to do their best and working many more project and focus areas than they were in 1998, with little increase in salary and staffing. Clearly, there are valid political drivers, as well as changes in threats and risks, but being pulled in multiple directions is a challenge for the emergency management community.

Terrorism or All Hazards

Terrorism versus all hazards continues to be a struggle. It often seems as if leaders at the Department of Homeland Security believe that no one at the state and local levels planned for terrorism prior to 9/11. The reality is FEMA, state emergency management agencies, and local emergency management agencies were developing anti-terrorist plans, conducting terrorist response and recovery training, and developing and participating in an active exercise program focused on terrorist attacks. Terrorism as a hazard that we had to plan for was not new.

Then we had 9/11 and the response to it was, "Throw everything out that you were doing before, because we (DHS) are going to do it a different way. Throw out the federal response plan because we need a national response plan. Throw out the way you were operating because now we need a national incident management system. Throw out a focus on all hazards because terrorism is different."

The reality is that if a building collapses, the response is not going to be that different. The follow-on investigation clearly will be different, and clearly there will be a need to protect evidence and ensure that we have the right people there to try to identify the perpetrators. But generally the response will look very similar whether the building collapses from a tornado or a bomb. I would suggest that terrorism is a hazard like hurricanes, like tornadoes, like floods and dam failures, like chemical incidents, like nuclear power plant incidents—in short, like all the events identified in a hazard and risk assessment. With a comprehensive, all-hazards plan, terrorism becomes a planning element, not a separate plan in and of itself. DHS espouses an all-hazards approach; yet it continues to put out plans, conduct exercises, and distribute grants that are focused on terrorism.

NIMS: Good Idea, Challenging Execution

The National Incident Management System (NIMS) is a personal issue for me. I think the Incident Command System (ICS) is an outstanding tool for first responders. But what DHS has done is taken a tool that works very well on a house fire or a building collapse or a terrorist attack on the Pentagon, and tried to make it fit every hazard, every scenario on a national basis.

In Ohio, we are, to the best of our ability, integrating NIMS, which is based on the incident command system, into all of our plans and procedures. ICS came out of the fire service. It was a system that ensures that lines of authority, command and control, resource allocation, and communications are fully addressed and understood by first responders reporting to a scene. It is tremendously valuable at the incident level; yet we are being asked to apply this concept to a flood that affects sixteen counties. Who is the incident commander in a countywide flood? There is not one spot or one point on the map where you can say, "There, that spot is the incident, and that fire chief is the incident commander." I see limited value in taking what was a tool for an incident with a defined location, and saying we will build a national incident management system based on what works very well at the local level.

There are smart people who would argue with me and fervently believe this is a great tool. They argue that it is building capability, and that while it is difficult right now, in ten years, it will all work, and we will be better prepared. I hope they are right. In the meantime, I will continue to struggle with a concept that is intended to be used in every event from a New Madrid earthquake that involves multiple states down to a hazardous materials incident and yet still refers to THE incident commander.

A purely logistical challenge I have with NIMS is the requirement for each state to certify every year that we meet certain implementation criteria to be eligible for all preparedness funding coming from the federal government. The federal assumption seems to be that since this is a national system, states are all implementing this in the same way. The reality is each state has a different culture and relationship with its local partners and will implement NIMS in different ways. Ohio is a home rule state, as described in the Ohio Local Government Structure and Finance Bulletin 835-98:

> All local governments in Ohio can be classified as either home rule or non-home rule, according to the authority given them by the Constitution and legislation in the *Ohio Revised Code.* All municipalities, and counties that have adopted a charter, are home rule. All other local governments are non-home rule. Non-home rule governments can only provide those services and perform those acts as specifically authorized by law. Home rule

> governments can perform all functions not specifically prohibited by law. Recent legislation permits townships, by vote of the people, to adopt a limited self-government form of government.

This generally means the lowest level of government has some level of autonomy. In spite of the fact that local jurisdictions have a fair amount of autonomy, I am required to certify that Ohio has met all of our NIMS compliance activities for this year—Ohio the state, not state agency—from the governor all the way down to that township trustee, all have completed all their requirements to ensure that we are NIMS-compliant. Relying on Internet tools and strong county directors, I have been able to certify that Ohio has made a good-faith effort to be NIMS-compliant, but requiring a state official to certify the actions of local government is, I believe, a flawed approach.

There are people in DHS who seem to think that every state has the authority to enforce activities like NIMS compliance down to the lowest level of government. This can be a big issue for a home rule state. It seems that some of our federal partners have lost sight of the nuances in states' cultures and capabilities that are part of the strength of our nation. We need to recognize these differences, encourage them when appropriate, and use them to build on the strengths unique to our states and regions. Instead, the prevailing philosophy seems to be that we must all look alike, plan alike, and respond alike.

National versus Local

There is this desire to build everything on a national level. We must have a national response plan and a national incident management system so everyone responds the same. The reality is that the response happens first at the local level.

In Ohio, I have certain planning criteria that we ask counties to use so that they have addressed the planning issues we've identified in Ohio. I don't tell them what their plan should look like, just what is must encompass. It is evaluated and must meet basic planning guidance. Ultimately, I believe the EMA director at the county level best knows what his or her threats are, and what capabilities they should be building to meet their county's needs.

I believe the same of states and their ability to identify and address their critical issues. Guidance is always valuable, and new threats and capabilities require flexibility, but a national, one-size-fits-all plan does a disservice to the diversity of our great nation.

Expectations

I think we have created a culture of entitlement. In many cases, our citizens expect the government to protect them from every hazard and, if something should occur, make them whole. There does not, however, seem to be a willingness to protect oneself, either from the physical effects of a hazard or from the financial impacts. It is a dichotomy we face in every disaster. People don't evacuate when they are told to, find themselves on their roofs surrounded by water, and then begin wailing for someone to rescue them. I want to respond by asking them, "What did you do when you were told to evacuate?"

There are certainly those who, because of special needs or lack of transportation, do not have the means to evacuate. It is clearly the responsibility of state and local government to plan for the effective evacuation of those populations. But many times, the people we see on the nightly news are those who had the means to evacuate, but chose not to leave. Their decision to ignore evacuation orders, at best, put first responders in harm's way and, at worst, leads to their own death. The media will find and focus on these victims, often asking why the federal or state response is so slow. I seldom hear them ask the victim why they didn't leave when asked to; why they didn't follow the guidance of their local and state officials; why they didn't prepare themselves and their families for the hurricane, the flood, the tornado. These are questions we need to begin to ask.

The sense of entitlement extends well into recovery programs. In fact, that is where it is probably strongest. Recently, I received a complaint from an individual in Ohio affected by a flood. The individual had applied for a grant to repair their apartment, which was destroyed, but as a renter, was not eligible for a repair grant. Unfortunately, the owner of the property wasn't interested in repairing the property. The applicant was angry because FEMA and the state weren't solving the problem. The individual was living in a hotel, not willing to look for another apartment, set on moving back into the destroyed property. Their complaint was, "Where is FEMA? Where is the state? Why isn't someone taking care of me?" Yet what is that individual doing to help themselves? Since there were other available properties in the area, the renter had multiple options to move to safe and sanitary housing, but wasn't willing to accept that solution.

We have created that culture of "Fix me; make me whole; it is not my responsibility to do anything to prepare." I do not know how we get around that. Part of that is probably a function of being more urban in terms of environment. I have a garden, but it is strictly those things that I want to grow because I like to have a garden. My father-in-law, who was a child of the Depression, had a garden to feed his family. So I think some of that is a shift of overall culture in the United States, but a large portion of it is expectation management.

Expectation management has suffered as FEMA "leans forward." They have an understandable attitude of "Never again will FEMA go through a Katrina." I don't blame them; they got crucified. I know they were doing much more than the media reported, and much more than leadership at the federal level acknowledged. The unfortunate result, though, is we have gone from a focus in the 1990s on managing expectations of our constituents to creating expectations of immediate response and recovery.

We've lost the expectation that residents be prepared to be self-sustaining for seventy-two hours. FEMA no longer waits until states request a liaison. They now plan to deploy at the initial stages of an event. ("Never again will FEMA go through a Katrina.") A FEMA presence either at a state EOC (emergency operations center) or at the scene of a disaster creates the expectation that a declaration is on the way. Why else would they be there? If that assistance isn't forthcoming or takes too long, the perception is that "Something went wrong. Somebody screwed up, and we're not getting the help we deserve." So I am concerned that this overwhelming drive to lean forward is blowing expectations out of the water. Expectation management is not an issue I hear FEMA talking about anymore.

I do not know how we get back to a culture of self-sufficiency. We talk about it, but we just cannot seem to get there. To some degree, we must understand the very basic economic drivers of those who cannot care for themselves, versus those who choose not to. It is not a valid argument for me, since I have the means, to look at a single mother working two jobs and say, "You need to prepare for pan flu and have a seven-day food and water supply in your house." It is unrealistic and, to some degree, not within their means to do. It comes down to those decisions that people struggle to make: Do I feed my kids today or stock food for a maybe event tomorrow? Do I stay home when I am sick to prevent spreading the illness or pay my rent next month? These are very real issues for a large segment of our population.

Much effort has gone into encouraging our citizens to prepare themselves and their families for disasters. From the Civil Defense days of "duck and cover" and bomb shelters to the preparedness kits and focus on self-sustainment for seventy-two hours in the 1990s to duct tape and a "culture of preparedness" in the new millennium.

The best efforts begin with our children and become an accepted part of their daily lives. The intensive effort on the part of the fire community has resulted in a generation that knows exactly what to do if their clothes catch on fire. Ask any kindergartner: "Stop, drop, and roll." Somehow, we need to incorporate these kinds of key messages into our daily language. We need to move people from saying they have a disaster plan "in my head" to writing it down on paper and then practicing it. Consistency and patience are key. People won't make major changes in their levels of personal preparedness unless there is a compelling, personal interest in doing so—or it becomes an accepted part of daily life.

Considerations

The reality is that finances have an impact on disaster relief efforts. Ohio's current budget situation is challenging. Those fiscal constraints may drive difficult decisions on our participation in disaster recovery programs. When a state receives a major disaster declaration for public assistance, there are some options on who pays the non-federal share. Historically, if Ohio has a public assistance disaster declaration, the state picks up 12.5 percent of the non-federal share for local governments, and the local applicant picks up the other 12.5 percent. We do not have a separate disaster fund, so the state's share comes out of the general revenue. Some states have specific set-aside funds; Ohio does not. We have to identify the projected cost and then go to our controlling board and the governor's office and request that the funds be set aside.

Major disaster declarations for individual assistance have a 25 percent non-federal share that is the responsibility of the state. We use the same process to acquire the non-federal share of IA declarations.

FEMA is responsive: All of these points are arguments, issues, and concerns that FEMA is aware of and, in some case, trying to address. The FEMA administrator, Dave Paulison, has really tried, knowing that FEMA will lean forward, to make sure they are doing it in a way that has minimal negative impact on states and locals. I do think there is a sincere attempt to hear the issues, understand them, and if appropriate, change. Certainly there are times when the answer is no. There are times we will not agree. NIMS comes to mind. I may not like it, but I do believe they are hearing our issues.

I think FEMA has made a much more concerted effort to engage states and locals early in the process of handling the ongoing changes in natural disaster relief. There are a variety of ways they are reaching out or through legislation have developed counsels that can provide this input up through the channels. At the state level, I think most directors believe we can pick up the phone and call our regional administrator or senior leaders at FEMA headquarters. FEMA has been far more open and responsive to states' concerns.

Of course, sometimes it comes down to "We will agree to disagree." There may be times that I may say, "I hear what you are saying, FEMA, but if it is not required by law, then that is not how we are going to do it in Ohio." The bottom line is that this is my state, and these are my governor's constituents, and I first and foremost serve my governor and the people of Ohio. At the end of the day, I am hopeful that my service supports not only my efforts here in Ohio, but my federal partners, as well.

As executive director of the Ohio Emergency Management Agency, Nancy J. Dragani leads the state's coordinated preparedness and response to both manmade and natural disasters, administers the State Homeland Security Funding Program, and oversees disaster recovery and mitigation efforts. Ms. Dragani was reappointed to her position by Governor Ted Strickland in 2007. She has been the director of the Ohio Emergency Management Agency since January 2005.

In September 2007, Ms. Dragani was elected vice president of the National Emergency Management Association; she will become the president of the national association in September 2008. Ms. Dragani is one of two state emergency management directors on the Federal Emergency Management Agency's National Advisory Council.

Prior to becoming the director, Ms. Dragani served as the director of operations at Ohio EMA, where she was responsible for emergency preparedness training, exercises, planning, homeland security and preparedness grants, and response operations in the State Emergency Operations Center during disasters.

During her career, Ms. Dragani has held a variety of positions with local, state, and federal government, including the director of administration and public information chief for the Ohio EMA, editor of the Ohio National Guard's quarterly magazine, and radio and television broadcaster with the U.S. Army. Ms. Dragani retired from the Ohio National Guard with twenty-two years of combined U.S. Army, Army National Guard, and Air National Guard service.

Ms. Dragani graduated summa cum laude from Ohio Dominican College, with a Bachelor of Arts degree, emphasis in communications.